Lecture Notes in Computer Science 8883

Commenced Publication in 1973
Founding and Former Series Editors:
Gerhard Goos, Juris Hartmanis, and Jan van Leeuwen

Srinath Srinivasa   Sameep Mehta (Eds.)

Big Data Analytics

Third International Conference, BDA 2014
New Delhi, India, December 20-23, 2014
Proceedings

 Springer

Volume Editors

Srinath Srinivasa
International Institute of Information Technology - Bangalore
26/C, Electronics City
Hosur Road
Bangalore - 560100, India
E-mail: sri@iiitb.ac.in

Sameep Mehta
IBM Research - India
4 Block C
Institutional Area
Vasant Kunj
New Delhi - 110070, India
E-mail: sameepmehta@in.ibm.com

ISSN 0302-9743   e-ISSN 1611-3349
ISBN 978-3-319-13819-0   e-ISBN 978-3-319-13820-6
DOI 10.1007/978-3-319-13820-6
Springer Cham Heidelberg New York Dordrecht London

Library of Congress Control Number: 2014956260

LNCS Sublibrary: SL 3 – Information Systems and Application,
incl. Internet/Web and HCI

Typesetting: Camera-ready by author, data conversion by Scientific Publishing Services, Chennai, India

Printed on acid-free paper

Springer is part of Springer Science+Business Media (www.springer.com)

Preface

This volume contains the papers presented at BDA 2014: The Third International Conference on Big Data Analytics, held during December 20–23, 2014 in New Delhi.

There were 35 submissions. Each submission received an average of 1.7 reviews from which the committee decided to accept 17 papers.

BDA 2014 was the third in this conference series that began in 2012. The aim of this conference is to enourage dissemination of results from research and practice involving very large datasets.

Given the data deluge from current-day technologies, the idea of Big Data is catching the interest of corporations and governments worldwide. This increased interest is also bringing with it a certain level of hype as well as skepticism about the promise of Big Data. Against this backdrop, such a conference assumes even more significance as it aims to showcase conceptual challenges and opportunities brought about by the data deluge.

The program was divided into six thematic sessions in addition to three keynote addresses, five tutorials and two invited talks.

The session on "Media Analytics" comprised three papers – one on enriching news articles with images from Instagram, and the other two on analyzing Twitter feeds for sentiment analysis and predicting election results.

The second thematic session was on "GeoSpatial Big Data," comprising four papers. Mamatha et al., analyze 30 years of weather data to look for similarity patterns in daily and weekly weather conditions. Suzuki et al. analyze sensor data from the Kaguya Moon Mission to identify geographic entities. Kannan et al. propose a catalyzation of model discovery in environmental datasets by augmenting them with semantics. Gupta and Lakshminarasimhan propose a MapReduce architecture for processing of spatio-temporal data for supporting analytical queries.

The third thematic session was on "Semantics and Data Models." It comprised two papers. Tripathi and Banerjee propose an optimizer for SPARQL queries suitable for Big Data queries. Batra and Sachdeva explore data models for managing electronic health records (EHRs).

The next thematic session was on "Search and Retrieval" and included of two papers by the same set of authors. In their first paper, Veningston and Shanmugalakshmi use techniques from ant colony optimization and a term association graph for query reformulation to provide personalized search results on Web search engines. Their second paper is also on personalization, and proposes a probabilistic model to rank the relevance of search engine results to a user based on their search profiles.

This was followed by the session on "Graphics and Visualization" – also made up of two papers. Kumari et al., present their work on a parallel algorithm on structural feature extraction from LiDAR point cloud datasets. Their work also addresses the challenge of remotely rendering such structural features, by efficiently transmitting only the required elements on a network. Gupta et al., propose a Big Data stack for visual analytics. They also address integration of different sources of data through non-planar graph-structured semantics. Graph analysis algorithms are also presented that can extract specific semantics from this aggregated dataset.

The last thematic session was on "Application-Specific Big Data," comprising three papers. Paul et al., present their work on dynamic routing of jobs in a multi-site distributed environment. Singh and Bansal present their industrial solution offering analytics-based interventions for optimization of energy use. Kumar et al. address datasets pertaining to protein–protein interaction networks and present an algorithm for effective function prediction.

In addition to the above, the proceedings also contain a tutorial paper on analytics-based approaches detecting fraud and money laundering.

The editors would like to acknowledge the untiring efforts of the Steering and Organizing Committee, as well as the rich participation from the Program Committee. Special thanks are due to our sponsors IBM Research India, Xerox India Research Centre, and NIXI, Government of India. Finally, the entire operation of managing submissions and creating these proceedings would have been much more tedious if it were not for the silently efficient EasyChair (www.easychair.org) conference management system, on which this preface too was written.

October 2014   Srinath Srinivasa
                                                                    Sameep Mehta

Organization

Program Committee

R.K. Agrawal	Jawaharlal Nehru University, New Delhi, India
Avishek Anand	Max Plank Institute, Germany
Subhash Bhalla	University of Aizu, Japan
Raj Bhatnagar	University of Cincinnati, USA
Vasudha Bhatnagar	Delhi University, India
Arnab Bhattacharya	IIT Kanpur, India
Janakiram D.	IIT Madras, India
Prasad Deshpande	IBM Research, India
Lipika Dey	TCS Innovation Lab, Delhi, India
Shady Elbassuoni	American University of Beirut, Lebanon
Vikram Goyal	IIIT Delhi, India
Rajeev Gupta	IBM Research India
Shyam Gupta	IIIT Delhi, India
Sharanjit Kaur	Delhi University, India
Pramod Kompalli	IIIT Hyderabad, India
Akhil Kumar	Penn State University, USA
Naveen Kumar	Delhi University, India
Choudur Lakshminarayan	Hewlett-Packard Laboratories, USA
Ravi Madipadaga	Carl Zeiss Labs, India
Geetha Manjunath	Xerox Research Centre, India
Sameep Mehta	IBM Research India
Mukesh Mohania	IBM Research India
Yasuhiko Morimoto	Hiroshima University, Japan
Mandar Mutalikdesai	Dataweave Solutions, India
Ullas Nambiar	EMC India
Hariprasad Nellitheertha	Intel Technology India Private Ltd., India
Anjaneyulu Pasala	Infosys Ltd., India
Jyoti Pawar	Goa University, India
Lukas Pichl	International Christian University, Japan
Emmanuel S. Pilli	Malaviya National Institute of Technology, India
Krishna Reddy Polepalli	IIIT Hyderabad, India
Sriram Raghavan	IBM Research India
Santhanagopalan Rajagopalan	IIIT Bangalore, India

Maya Ramanath   IIIT Delhi, India
Chandrashekar Ramanathan   IIIT, Bangalore
Sayan Ranu   IIT Madras, India
Shourya Roy   Xerox Research Centre, India
Mark Sifer   University of Wollongong, New Zealand
Srinath Srinivasa   IIIT Bangalore, India
Saurabh Srivastava   IBM Research India
Lv Subramaniam   IBM Research India
Shamik Sural   IIT Kharagpur, India
Ashish Sureka   IIIT Delhi, India
Srikanta Tirthapura   Iowa State University, USA

Additional Reviewers

Agrawal, Swati   Kulkarni, Sumant
Asnani, Kavita   Madaan, Aastha
Gupta, Shikha

Table of Contents

A Framework to Improve Reuse
in Weather-Based Decision Support Systems

A. Mamatha[1], Polepalli Krishna Reddy[1], Mittapally Kumara Swamy[1],
G. Sreenivas[2], and D. Raji Reddy[2]

[1] International Institute of Information Technology-Hyderabad (IIIT-H),
Telangana State, India
[2] Professor K. Jayashankar Telangana State Agricultural University,
Hyderabad, India
{mamatha.a,kumaraswamy}@research.iiit.ac.in, pkreddy@iiit.ac.in,
{gsreenivas2002,dandareddy009}@gmail.com

Abstract. The systems for weather observation and forecast are being operated to deal with adverse weather in general to mankind. Weather-based decision support systems (DSSs) are being build to improve the efficiency of the production systems in the domains of health, agriculture, livestock, transport, business, planing, governance and so on. The weather-based DSS provides appropriate suggestions based on the weather condition of the given period for the selected domain. In the literature, the notion of *reuse* is being employed in improving the efficiency of DSSs. In this paper, we have proposed a framework to identify similar weather conditions, which could help in improving the performance of weather-based DSSs with better *reuse*. In the proposed framework, the range of weather variable is divided into categories based on its influence on that domain. We form a weather condition for a period which is the combination of category values of weather variables. By comparing the daily/weekly weather conditions of a given year to weather conditions of subsequent years, the proposed framework identifies the extent of *reuse*. We have conducted the experiment by applying the proposed framework on 30 years of weather data of Rajendranagar, Hyderabad and using the categories employed by India Meteorological Department in Meteorology domain. The results show that there is a significant degree of similarity among daily and weekly weather conditions over the years. The results provide an opportunity to improve the efficiency of weather-based DSSs by improving the degree of *reuse* of the developed suggestions/knowledge for the corresponding weather conditions.

Keywords: Reuse, Decision Support Systems, Similarity, Weather Condition, Data Analysis.

1 Introduction

Weather is the state of the atmosphere as measured on a scale of hot or cold, wet or dry, calm or storm, and clear or cloudy [1, 2]. Almost all occupations of

S. Srinivasa and S. Mehta (Eds.): BDA 2014, LNCS 8883, pp. 1–13, 2014.

life are influenced by weather. Proper use of weather information can reduce the risk. Also, proper understanding/application of observed and forecast weather information/alerts will improve the efficiency of planning and execution of tasks in all walks of life, occupations and services.

The systems for weather observation and forecast are being operated to deal with adverse weather in general to the mankind. The weather observation and forecasting systems have become vital for every country as they deal with weather and its related factors affecting production systems. Governments are investing huge budgets and employing advanced computer information systems for weather forecast. Over the years, the weather information and forewarning systems are gradually becoming powerful, domain-specific and location-specific. The weather observatory and forecast systems return the values of weather parameters like rain fall, temperature, cloud cover, humidity, wind speed and wind direction.

The Decision Support System (DSS) is an interactive computer-based information system that supports decisions in various domains. Weather-based decision support systems (DSSs) are being build to improve the efficiency of the production systems in the domains of health, agriculture, livestock, transport, business, planning, governance and so on. Based on weather condition, the weather-based DSS provide appropriate suggestions to the stakeholder.

The notion of *reuse* is one of the important opportunity to improve the performance of DSS. In this paper, we have proposed a framework to find similar weather conditions among weather data over the years which could help in improving the performance of weather-based DSSs by improving the *reuse*. In the proposed framework, the range of weather variable is divided into categories based on its influence in the respective domain. We define a notion of *weather condition for a duration* which is the combination of category values of statistics of weather variables for the given duration. By comparing the weather conditions of the given year to subsequent years, the proposed framework identifies similar weather conditions. We have applied the proposed framework on 30 years of weather data of weather station at Rajendranagar, Hyderabad, India by using the categories employed by India Meteorological Department (IMD) in Meteorology domain. The experiment results show that there is a significant degree of similarity among daily and weekly weather conditions over the years. The proposed approach provides a scope to improve the efficiency of weather-based DSSs by maximizing *reuse* of the developed content (such as standardized suggestions/content/advisory) for the relatively lesser number of weather conditions.

The organization of paper is as follows. In the next section, we present the related work. In section 3, we explain the proposed approach to extract similar weather conditions. In section 4, we present the case study. The last section contains summary and conclusion.

2   Related Work

The notion of reuse is widely employed to improve the performance of software development processes and efficiency of large scale systems [3]. The *reuse* improves performance of software project management [4]. Software product lines (SPL) refers to software engineering methods, tools and techniques for creating a collection of similar software systems from a shared set of software assets using a common means of production. An approach for domain-oriented geographical SPL is explained with a case study of marine ecology sub-domain [5].

The DSSs solve many semi-structured and unstructured problems and help to make decisions which may change rapidly. The utilization and technology issues in the context of large scale complex system in DSSs are explained in [10]. DSSs play a key role for medical diagnosis [11]. The capture and reuse of ICT-based knowledge in clinical decision support system (CDSS) has been explained in [12]. An approach for knowledge reuse for growth-assessment of children is discussed in [13].

The dependency on weather is expanding to many domains. An approach to investigate the impact of weather on travel time prediction is explained in [14]. A DSS [15] is built to control fungal diseases in winter wheat crop. An agro-advisory system (AAS) to reduce investment and loss is explained in [16]. Reuse in agro-advisories is propose using the notion of *weather window* for dominant crops [6]. A framework is proposed to improve the practical agricultural education using the notion virtual crop labs [9]. It was reported in [7] that the performance of subject matter specialist in IT-based agro-meteorological system could be improved by extracting similar advice. In [8], a frame work to reuse the content is proposed to improve the performance of agro-advisory system.

In agriculture-based DSSs, decisions can be improved with accurate weather forecast. A system for efficient crop management based on improved weather forecast is proposed in [17]. The importance of a web-based DSS is proposed for minimizing climate risks in agriculture is discussed in [18].

In this paper, we have proposed a framework for weather-based DSSs for analyzing the similarity among daily and weekly weather conditions based on domain specific categories.

3   Proposed Approach

The weather is represented through the values of weather variables. The examples of weather variables are rain fall (RF), maximum temperature (Tmax), minimum temperature (Tmin), maximum relative humidity (maxRH)and minimum relative humidity (minRH). The units of RF, temperature, humidity are millimeter (mm), degree centigrade (deg C), percent (%) respectively. The sample weather values for 14 days of Rajendranagar weather station for 2009 year are given in Table 1.

Note that there are several other weather variables like fog, air pressure and so on. In this paper, we explain the proposed approach by considering preceding weather variables. However, the proposed approach is a general approach and applicable for other variables.

Table 1. Sample daily weather data from 1st Jan 2009 to 14th Jan 2009, collected at Rajendranagar weather station, Hyderabad, Telangana State

Date	RF	Tmax	Tmin	RHmax	RHmin
1-Jan	0.0	29.5	13.5	90.0	32.0
2-Jan	0.0	28.5	13.5	89.0	34.0
3-Jan	0.0	28.5	15.5	90.0	39.0
4-Jan	0.0	28.0	13.2	93.0	35.0
5-Jan	0.0	28.5	13.0	96.0	35.0
6-Jan	0.0	28.5	14.0	90.0	36.0
7-Jan	0.0	29.0	14.5	81.0	88.0
8-Jan	0.0	28.5	16.5	91.0	51.0
9-Jan	0.0	27.5	19.0	82.0	51.0
10-Jan	0.0	28.0	18.5	91.0	51.0
11-Jan	0.0	28.0	16.0	96.0	40.0
12-Jan	0.0	29.5	15.0	90.0	41.0
13-Jan	0.0	29.5	14.0	92.0	34.0
14-Jan	0.0	27.5	13.5	87.0	34.0

The problem statement is as follows. Given daily weather values of weather variables over year, how to calculate the extent of similar weather in sub-sequent years?

Normally, the weather is compared after calculating weather statistics over certain duration of interest. The duration is fixed based on the application in the respective domain. Here, we define the notion of *weather condition for a duration d*.

Definition 1. *Weather Condition (WC(d, s, e))*. *Given a set of weather variables, the statistics of each weather variable calculated for a duration d is called WC(d, s, e). Here, s represents the start date, e represents the end date and d represents the number of days between s to e (both are inclusive)* .

The number of WCs for a year comes to $\lfloor \frac{365}{d} \rfloor$. When $d=1$, WC indicates the daily weather values of weather variables $\langle$ RF, Tmax, Tmin, maxRH, minRH, $\rangle$. However when d >1, the statistics for each weather parameter the weather variables $\langle$ RF, Tmax, Tmin, maxRH, minRH $\rangle$ are to be calculated. Note that appropriate function should be employed to compute the statistic parameter value. For Tmax, Tmin, maxRH and minRH the value is equal to the mean value over d days whereas for RF, the value represents the cumulative value over the d.

From Figure 1, WC(1, 1 Jan 2009, 1 Jan 2009) is given as $\langle 0.0, 29.5, 13.5,$ 90.0, 32.0$\rangle$ and WC(7, 1 Jan 2009, 7 Jan 2007) is given as $\langle 0.0, 28.6, 14.0, 89.9,$ 42.7$\rangle$. By using the notion of weather condition, the problem statement is as follows. Given d and corresponding WCs for each year, how to extract similar WCs from the weather data of subsequent years?

Definition 2. *Similar Weather Conditions.* *Let* i *and* j *be the identifiers of two weather conditions of different years or the same year with* p *weather variables over the same duration* d, *i.e.,* i = WC(d, *, *) *and* j = WC(d, *, *). *Let sim(i, j) indicate the similarity of weather conditions. We say, sim(i, j) is equal to 1, if the values of corresponding weather variables are equal. Otherwise, sim(i, j) is equal to 0.*

It can be observed that having real values of weather parameters, it is difficult to get high similarity among weather conditions as similarity value equals to zero even if one of the value differs by a small value. However, it can be observed that in medical domain or agrometeorology domain, a different suggestion or advice is not recommended for a small change, like 0.2 degree centigrade, in temperature value, or small change, like 2 per cent, in humidity value.

Based on preceding observation, we introduce a notion of category to improve the similarity among weather conditions. We divide the range of weather variable into different categories based on the influence in the application area or domain. Note that the categories of a weather variable may differ from domain to domain. For example, the convenient temperature range for human being is different than the convenient temperature range for livestock or crop. For each weather variable, we divide the range of that weather variable into different classes. Each class is termed as a category. The category is the description/name of that class. For example, the domain of relative humidity can be divided into: LOW, MODERATE, HIGH, VERY HIGH and so on. Each other weather parameter is divided to different categories based on the influence on the respective domain. For example, suppose if the relative humidity range from 31% to 60 % categorized as MODERATE and let the mean value of relative humidity over 7 days is equal to 32 degree centigrade. We consider the RHmin=32 as RHmin=MODERATE. Similarly, the values of other variables are mapped to the corresponding categories. Given the categories of each weather variable, we now define the term *category-based weather condition* to represent weather condition for duration 'd'.

Definition 3. *Category-based Weather Condition (CWC(d, s, e)). The notation* d,s,e *represent the same notion as WC. Let the ⟨range, category⟩ table contains the details of range and category names for each weather variable. Given WC(d, s, e), we get CWC(d, s, e), if we replace the values of each weather variable with the corresponding categories by referring the table ⟨range, category⟩.*

By using the notion of CWC, the problem statement is as follows. Given *'d'* and corresponding CWCs for each year, how to extract similar CWCs from the weather data of subsequent years?

Definition 4. *Similarity of CWC* *Let* i *and* j *be the identifiers of two CWC of different years or the same year with* p *weather variables over the same duration* d, *i.e.,* i = CWC(d, *, *) *and* j = CWC(d, *, *). *Let sim(i, j) indicate the similarity of category-based weather conditions. We say, sim(i, j) is equal to 1, if the categorical values of corresponding weather variables are equal. Otherwise, sim(i, j) is equal to 0.*

Proposed Approach. Given a domain, the proposed approach to find the similar weather conditions is as follows. At first, for the given domain and application, we obtain the categories defined for each weather variable based on the effect of each weather variable on the stakeholder of the application. The values of duration are also decided based on the utility of weather condition-based suggestions for the stakeholder. After forming the CWC for each year of weather data, we can extract similar WCs by comparing the WCs of the given year to subsequent years.

Performance metric We define the performance metric to measure the extent of similar WCs of given year with reference to preceding years. That is, given a year x we would like to calculate the percentage of WCs of x which are similar to WCs of preceding n years ($n \geq 1$).

Definition 5. *Coverage Percentage of weather conditions of year* x *over preceding* n *years (CP(x/n)). Given weather conditions of* x *and* n *preceding years of years* x *, CP(x/n) is given by the percentage of weather conditions of* x *which appear in the set of WCs of preceding* n *years.*

The pseudo code of the proposed approach is given in Algorithm 1. The input to this algorithm is weather data and categories of weather variables defined for the corresponding domain. Note that the categories are domain-specific and are decided by subject matter specialists or experts in the respective domain based on the applicability and utility for the intended application. Also, the duration is also given as an input. Based on the duration, the approach takes weather data over n years as an input and generated weather conditions. Based on the categories, it generates categorical weather conditions. Next, for each year x, it computes $CP(x/n)$.

Algorithm 1. Computing Coverage Percentage of Weather Condition

```
1: INPUT n, p, weather data, categories
2: //Weather data of p variables for n years
3: //Categories of each variable of the form < range, category >
4: OUTPUT CP for each year i, i in 1..n.
5: begin
6: WC(n), CWC(n) represents array of sets of WCs.
7: S represents set of WCs of i-1 years ; d represents duration.
8: for i in 1.. n do
9:      Compute WCs for year i;
10:     Compute CWCs for year i;
11: end for
12: initialize S = CWC(1);
13: for i in 2.. n do
14:     CP(i/(i-1))=|CWC(i)∩S| / |CWC(i)|;
15:     S = S ∪ CWC(i);
16: end for
17: end
```

4   Case Study

We evaluate the performance of the proposed approach by considering the case study in meteorology domain. IMD is an agency of the Ministry of Earth Sciences of the Government of India [19]. It is the principal agency responsible for meteorological observations, weather forecasting and seismology. IMD has started weather services for farmers in the year 1945. Currently, IMD is issuing weather summaries for the following: now casting (in which the details about the current weather and forecasts up to a few hours ahead are given), short range forecasts (1 to 3 days), medium range forecasts (4 to 10 days) and long range/extended range forecasts (more than 10 days to a season). As a part of medium range forecast, from 1st June 2008, IMD has started issuing quantitative district level (612 districts) weather forecast up to five days, twice a week.

The details of data set and categories used in the experiment are as follows.

- **Weather Data:** We collected 30 years of weather data of Rajendranagar, Hyderabad from 1980-2009 each year consists of daily weather conditions of currently year i.e., 365 weather conditions for non leap year and 366 weather conditions for leap year each with 5 parameters: RF, Tmax, Tmin, maxRH, and minRH.
- **Categories of weather variables:** India Meteorological Department (IMD) prepares weather summaries about forecast. Based on the influence of weather parameters on the domain, IMD has assigned categories for specified range of each of these parameters which are called categories which are presented in Table 2 [19]. These categories are used to prepare weather summaries for forecast data. The categories of RF(Rain Fall) and RH (Relative Humidity) are divided into ranges. The categories of temperature are defined based on deviation from normal. Climatologists understand the trends of both forecast and observed data using the corresponding climatic normals. A climatic normal is defined as the arithmetic average of a climate element such as temperature over a prescribed 30-year interval. So, the categories for temperature variable are to be assigned by referring Table 2 which are based on climatic normals.

Methodology: We have taken 30 year weather data and calculated CWC by considering *duration* = 1 and *duration* = 7. The duration as one day indicates the case of short range weather forecast and duration as 7 days indicates the case of for medium range weather forecast provided by IMD. The season-based WCs are calculated by dividing the year in to three seasons namely *Summer (March to May), Kharif (June to November), Rabi (December to March)*.

- Coverage of WCs with duration = 1 day
- Coverage of season-based WCs with duration = 1 day
- Coverage of WCs with duration = 7 days
- Coverage of season-based WCs with duration = 7 days.

In these graphs we represent year number on the x-axis. The year number "1" represents 1980. For each year, we plotted coverage percentage of respective WCs of that year similar to the WCs of preceding years. For example, the CP(3/2) equal to the number of WCs in 1982 which similar to the WCs set of 1980 and 1981. Similarly, CP(20/19) equal to number of WCs in the year 1999 which are similar to the WCs set of the years 1980 to 1998.

Experiment 1. Weather conditions with duration = 1 day: We have have carried out similarity analysis for numeric WCs and CWCs. The total WCs comes to $10,958 (= 22 * 365 + 8 * 366)$ weather conditions. For CWCs, the results plotted in Figure 1 show that the value of CP(2/1), CP(4/3) and CP(6/5) as 86%, 94% and 97% respectively. The graph also shows the curve for CP for numeric WCs. It can be observed that the CP of numeric WCs is significantly lower than the CP of CWCs.

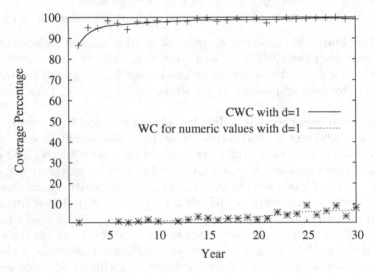

Fig. 1. Comparison of weather condition with d=1. Here, CWC represents category based weather condition, WC represents weather condition for numeric values.

Experiment 2. Season-based weather conditions with duration = 1 day: A weather condition in one season is different from the weather condition in different season. We have divided the year into three crop seasons: Summer, Kharif and Rabi. Next, we have compared the CWCs of the given season to the CWCs of the same seasons of the preceding years. The results plotted in Figure 2 show the value of CP(2/1), CP(4/3) and CP(6/5) for total year as 75%, 91% and 96% respectively. It can be observed that the CP values for numeric WCs is significantly lower than the CP values for CWCs.

Experiment 3. Weather Conditions with duration = 7 days: We have carried out similarity analysis of by computing WCs for 30 years i.e., $1,560$ $(= 30 * 52)$ with duration = 7 days. WC with duration = 7 days is obtained from daily weather values by calculating the mean for parameters like $\langle$ Tmax, Tmin,

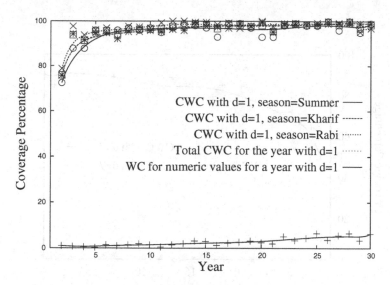

Fig. 2. Comparison of season-based weather condition with d=1. Here, CWC represents category based weather condition, WC represents weather condition for numeric values.

maxRH, minRH⟩ and cumulative of RF for each week. Appropriate weather tags are assigned to each weather value. For CWCs, the results plotted in Figure 3 show that the value of CP(2/1), CP(4/3) and CP(6/5) as 49%, 67% and 71% respectively. It can be observed that even though the CP performance is reduced as compared to the CP performance of CWCs with d=1, there is a significant values for CP (=67%) can be observed even in the fourth year. This indicates that there is a significant scope for performance improvement by exploiting reuse. As expected, the CP performance of numeric WC is significantly low.

Experiment 4: Season-based Weather Conditions with duration = 7 days: We have calculated the coverage percentage of weather conditions of current year based on season with the corresponding season-based weather conditions of previous years. For CWCs, the results in Figure 4 show that the value of CWC(2/1), CWC(4/3) and CWC(8/7) for total year as 45%, 64% and 75% respectively. From the results, it can be observed that, there is a significant CP values even in fourth year for season-based weather conditions. The performance of numeric WCs is significantly low.

Summary: The summary of the results indicate that, for small (duration = 1) and medium durations (duration = 7), several weather conditions of a year are similar to the weather conditions of preceding years. If we develop the knowledge-bases, advices or content in the meteorology domain for the same weather conditions, the knowledge could be reused for subsequent years. As a result, the performance of weather-based DSS could be improved.

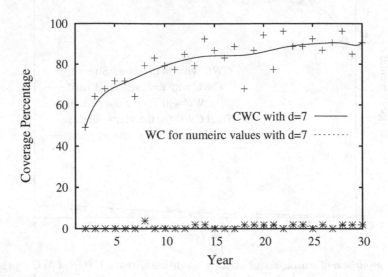

Fig. 3. Comparison of weather condition with d=7. Here, CWC represents category based weather condition, WC represents weather condition for numeric values.

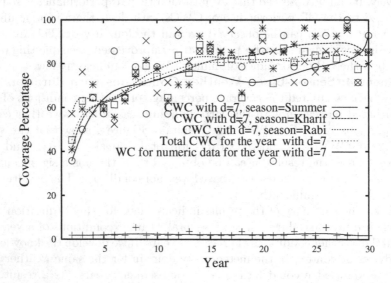

Fig. 4. Comparison of season-based weather condition with d=7. Here, CWC represents category based weather condition, WC represents weather condition for numeric values.

Table 2. Category defined by India Meteorological Department for weather variables

Weather Variable Name	Range	Description
Rain Fall (mm)	0 - 0	No Rain(NR)
	0.1 - 2.4	Very Light Rain(VLR)
	2.5 - 7.5	Light Rain(LR)
	7.6 - 35.5	Moderate Rain(MR)
	35.6 - 64.4	Rather Heavy Rain(RH)
	64.5 - 124.4	Heavy Rain(HR)
	124.4 - 244.4	Very Heavy Rain(VHR)
	>= 244.5	Extremely Heavy Rain(EHR)
Temperature(°C) deviation from normal	-1 ,0, 1	Little Change(LC)
	2 or -2	Rise/Fall(R/F)
	3 to 4	Appreciable Rise(AR)
	-3 to -4	Appreciable Fall(AF)
	5 to 6	Marked Rise(MR)
	−5 to -6	Marked Fall(MF)
	>= 7	Large Rise (LR)
	<= -7	Large Fall(LF)
Relative Humidity (%)	0 - 30	Low(L)
	31 - 60	Moderate(M)
	61 - 80	High(H)
	>= 81	Very High(VH)

5   Summary and Conclusion

Weather plays an important role in almost all aspects of the life. Hence, accurate and timely forecasting of weather has got wide implications ranging from increasing the agricultural production to reducing the damage to life and property.

In this paper, an effort has been made to propose a framework to compute the extent of similar weather situations. By defining the notion of weather condition for a given duration, we have proposed a framework to compute the similar weather conditions in the weather data over several years. We have also proposed the framework by defining the notion of weather condition by exploiting the categories defined for each weather variable in the respective domain.

We have reported a case study by extending the proposed framework to extract similar weather conditions in 30 years weather data of Rajendranagar observatory in Hyderabad and using the categories defined by IMD for preparing weather summaries. The results show that similarity among weather conditions of given year to other years is significantly high. The performance results on daily and weekly weather conditions show that there is a significant similarity among the weather conditions over the years. Also, as compared to the similarity over weather conditions based on real weather values, there is a significant increase in similar weather conditions based on categories. The results indicate that by defining domain-specific/application specific categories, it is possible to increase similarity among weather conditions over the years.

As we have analyzed the similarity among weather conditions by considering duration as one day and seven days, the case study can be utilized to identify the similar weather conditions of short range weather forecast and medium range weather forecast provided by IMD.

The proposed results can be used to build systems to improve the performance of weather-based DSS with better reuse. Note that, for certain domain, if we develop certain knowledge/content/suggestion for a particular weather condition, it can be reused whenever similar weather condition occurs in future.

The proposed framework provides several opportunities for future work. The notion of weather condition will be improved by capturing the change between the observed weather, and weather forecast. We will conduct more experiments by considering weather conditions of different durations and weather data at multiple locations across India. We will compute the coverage percentage of weather conditions by applying domain specific categories of weather variables such as health, crops, livestock and so on.

References

[1] Strahler, A.N.: Physical geography, 2nd edn., p. 185. John Wiley and Sons, New York (1960)

[2] Monkhouse, F.J.: A dictionary of geography. Edward Arnold (Publishers) Ltd, London (1978)

[3] Selby, R.W.: Enabling reuse-based sotware development of large-scale systems. IEEE Transactions on Software Engineering, 495–510 (2005)

[4] Srivastava, B.: A decision-support framework for component reuse and maintenance in software project management. Software Maintenance and Reengineering, 1534–5341 (2004)

[5] Buccella, A., Cechinch, A., Arias, M., Polla, M., Doldan, M.D.S., Morsan, E.: Towards systematic software reuse of GIS insights from a case study. Computers and Geosciences 54, 9–20 (2013)

[6] Balasubramanian, T.N., Jagannathan, R., Maratatham, N., Sathyamoorthi, K., Nagarajann, R.: Generation of weather window to develop agro advisories for Tamilnadu under automated weather forecast system. Journal of Agrometeorology 16(1), 60–68 (2014)

[7] Reddy, P.K., Trinath, A.V., Kumaraswamy, M., Reddy, B.B., Nagarani, K., Reddy, D.R., Sreenivas, G., Murthy, K.D., Rathore, L.S., Singh, K.K., Chattopadhyay, N.: Development of eAgromet prototype to improve the performance of integrated agromet advisory service. In: Madaan, A., Kikuchi, S., Bhalla, S. (eds.) DNIS 2014. LNCS, vol. 8381, pp. 168–188. Springer, Heidelberg (2014)

[8] Mahadevaiah, M., Raji Reddy, D., Sashikala, G., Sreenivas, G., Krishna Reddy, P., Bhaskar Redddy, B., Nagarani, K., Rathore, L.N., Singh, K.K., Chattopadhyay, N.: A framework to develop content for improving agromet advisories. In: The 8th Asian Federation for Information Technology in Agriculture (AFITA), Taipei (2012)

[9] Krishna Reddy, P., Bhaskar Reddy, B., Rama Rao, D.: A model of virtual crop labs as a cloud computing application for enhancing practical agricultural education. In: Srinivasa, S., Bhatnagar, V. (eds.) BDA 2012. LNCS, vol. 7678, pp. 62–76. Springer, Heidelberg (2012)

[10] Filip, F.G.: Decision support and control for large-scale complex systems. Annual Review in Control 32(1), 61–70 (2008)

[11] Michael Johnson, P., Zheng, K., Padman, R.: Modeling the longitudinality of user acceptance of technology with an evidence-adaptive clinical decision support system. Decision Support Systems, 57, 444–453 (2014)

[12] Andreea Ioana Stanescu, Florin Gheorghe FILIP:Capture and reuse of knowledge in ICT-based decisional environments. Informatica Economia, Vol. 13 (4) (2009)

[13] Kuilboer, M.M., Shahar, Y., Wilson, D.M., Musen, M.: Knowledge re-use, temporal abstraction mechanism for the assessment of children's growth. American Medical Informatics Association (1994)

[14] Bajwa, S.: Investigating the impact of weather information on travel time prediction. In: OPTTIMUM International symposium on recent advances in transport modelling (2013)

[15] Schepers, H.T.A.M., Bouma, E., Frahm, J., Volk, T., Secher, B.T.M.: Control of fungal diseases in winter wheat with appropriate dose rate and weather-based DSS. In: EPPP Conference on Forecasting, vol. 26, pp. 623–630 (1996)

[16] Maini, P., Rathore, L.S.: Economic impact assessment of the Agrometeorological Advisory Service of India. Current Science 101(10), 1296–1310 (2011)

[17] Jones, J.W., Hansen, J.W., Royce, F.S., Messina, C.D.: Potential benefits of climate forecasting to agriculture. Agriculture, Ecosystems and Environment 82, 169–184 (2000)

[18] Breuer, N.E., Cabrera, V.E., Ingram, K.T., Broad, K., Hildebrand, P.E.: AgClimate, A case study in participatory decision support system development. Climatic Change 87, 385–403 (2008)

[19] India Meteorological Department weather forecasters guide (August. 2014), http://www.imd.gov.in/section/nhac/dynamic/forecaster_guide.pdf

Suitability of Data Models
for Electronic Health Records Database

Shivani Batra and Shelly Sachdeva

Jaypee Institute of Information Technology University,
Department of Computer Science and Engineering,
Sector-128, 201301, Noida, India
ms.shivani.batra@gmail.com, shelly.sachdeva@jiit.ac.in

Abstract. With the advancement in technology, data is also growing exponentially. Storing this BIG DATA in an efficient manner is the key for any successful project. Work done in this paper is dedicated towards presenting the possible efficient ways available to store Electronic Health Records (EHRs). The main hurdles in storing EHRs are sparseness and volatility which relational model is incapable to handle. The other models present for storing EHRs are Entity Attribute Value (EAV), Dynamic Tables, Optimized Entity Attribute Value (OEAV) and Optimized Column Oriented Model (OCOM). Authors have provided a comparative study which will help the administrator to choose the best model among the models specified above. Authors have also discussed about the different scenarios (standardized and non-standardized EHRs) in which a combination of these models can be used. Authors have simulated EAV, Dynamic tables, OEAV and OCOM models to provide comparison results of time taken for executing basic operations (queries) and memory consumed by different models.

Keywords: Data Models, Electronic Health Records, Sparseness, Volatility, Heterogeneity, Entity Centric Query, Attribute Centric Query, Storage.

1   Introduction

"Big data is a blanket term for any collection of data sets so large and complex that it becomes difficult to process using on-hand data management tools or traditional data processing applications" [1]. Data exists everywhere (for example in schools/colleges, in hospitals, in industries, in banking, in internet, in astronomy and in finance). To manage these data we require data models. A data model defines a storage format for data. But just storing the big data is not enough; big data should be stored in a manner that it is easy for a user to capture, search, share, transfer, analyze and visualize the stored data [1]. When a data model achieves these goals (capture, search, share, transfer, analyze and visualize) most efficiently for a particular data set, it is said to be the best data model for that particular dataset. It is not necessary that a data model suitable for one type of data set is equivalently good for another dataset. For example, a school database consists of defined set of attributes and non-sparse instances, whereas healthcare database

S. Srinivasa and S. Mehta (Eds.): BDA 2014, LNCS 8883, pp. 14–32, 2014.
© Springer International Publishing Switzerland 2014

comprise of sparse data and new attributes can arrive anytime with the advancement in technology. Thus, relational model is best suitable for school database but not for healthcare database.

The current research focuses on electronic health records (EHRs) which contain multidimensional highly sparse data. Due to the presence of sparseness and volatility in healthcare data, relational model is not suitable [2-3]. Therefore, authors aims at providing a comparative study about all existing models which can give better results for capturing, searching, sharing, transferring, analyzing and visualizing healthcare data. Capturing, searching, sharing, transferring, analyzing and visualizing healthcare data speedily and efficiently is highly required because it can help in easy diagnosis of a patient disease and can even save the person life in critical scenarios.

In section 2 authors have provided details about EHRs. Section 3 inspects various models available to store EHRs. Section 4 provides the details of experiment performed to compare various models explained in section3. Section 5 finally presents the conclusions in this paper.

2   Electronic Health Records (EHRs)

Electronic Heath Records (EHRs) refers to the healthcare records which are stored electronically.. A human body consists of many parts and a human being can suffer from different types of diseases. Healthcare record of one person will contains all such health information about the person. When an administrator defines a structure (i.e. the set of attributes) of EHR; he considers about all the possible attributes which a patient can have. Further he chose to opt one of the following two cases:

2.1   Non – Standardized EHRs

Many organizations follow their own suitable set of attributes for EHRs as per their requirements. The set of attributes chosen vary from organization to organization i.e. there is no common standard followed by the organizations. Non-standardization leads to many problems such as interoperability.

2.2   Standardized EHRs

In case of non-standardized EHRs, administrator has complete liberty of choosing set of attributes as per his requirement whereas in case of standardized EHRs, set of attributes and terminology is predefined. The set of attributes in case of standardized EHRs is decided by one standard organization and adopted by all others. The organization mostly divides the set of attributes in two categories namely Mandatory attributes and Optional attributes. Mandatory attributes are the one which are constrained to have non null values whereas optional attributes can have null values.

A typical EHR contains six major categories – administration, nursing, clinical, laboratory, radiology and pharmacy as shown in fig. 1(a).

These categories have further defined set of attributes for example laboratory can contain attributes for blood pressure, body mass index, heart rate, diabetes and many more therefore EHR is very huge. There may be many cases where patient do not go for radiology or laboratory, in such cases the fields of EHR dedicated for laboratory and radiology will be kept null. This scenario happens often, so most of the fields in EHR are kept null which results in wastage of space. This is very huge in case of EHRs. So, the very first requirement for a data administrator of EHR is to manage the EHR such that minimum storage is used. Technically, this problem is addressed as sparseness.

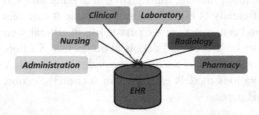

Fig. 1(a). Data from different departments of a hospital

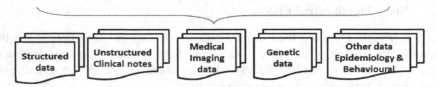

Fig. 1(b). EHR stored in a relational model

| Structured data | Unstructured Clinical notes | Medical Imaging data | Genetic data | Other data Epidemiology & Behavioural |

Fig. 1(c). Type of data stored in EHR

Relational model is the simplest way of modeling data (as shown in fig. 1(b)). Whenever a new concept arises in medical field, EHR needs to be modified to accommodate that concept. It might not be possible in relational model as relational model has a limit on maximum number of columns that can be defined. For example, the maximum number of columns in DB2 and Oracle is 1012 [4]. This problem is termed as volatility.

Modeling the data which is complex, sparse, interrelated, temporal and heterogeneous is a challenging task. EHR data explores and expands rapidly therefore, authors explained the standardized EHRs modeling techniques also. Standards have proposed a generic approach consisting of reference model and archetypes [5] which will be able to include the newly defined/modified data types.

The third requirement of EHR is heterogeneity i.e. capability of multiple data types. The EHRs data have multitude of representations [6]. As shown in fig. 1(c), the contents may be structured, semi-structured or unstructured, or a mixture of all three. These may be plain text, coded text, paragraphs, measured quantities (with values and units); date, time, date-time (and partial date/time); encapsulated data (multimedia, parsable content); basic types (such as Boolean, state variable); container types (list, set); and uniform resource identifiers (URI).

These three issues (sparseness, volatility and heterogeneity) laid the requirement of a new model. The most famous models which came into existence were Entity Attribute Value (EAV) model [7], Dynamic Tables [8], Optimized Entity Attribute Value (OEAV) [9] and Optimized Column Oriented Model (OCOM) [10]. The next section will give details about these models.

3 Data Models for EHRs Database

This research aims at doing a comparative analysis of various data models. EHRs database being sparse, volatile and heterogeneous requires an appropriate data model for facilitating its users with various database functionalities.

3.1 Entity Attribute Value Model

EAV model is a row based model which stores the value of an attribute in one row unlike relational model which stores whole attribute values in one dedicated column. EAV model stores data in row format in which each row consists of triplets Entity, Attribute and Value. 'Entity' part represents various entities of the dataset which can be used to uniquely identify the entity whose data is being stored, 'Attribute' part represents different column names corresponding to each entity i.e. the attribute name of the entity for which data value is stored in the particular row and 'Value' part represents the value of attribute for the particular entity.

Fig 2(a) represents a typical relational model and Fig 2(b) shows its equivalent EAV model. As shown in the Fig 2(a) and Fig 2(b), EAV model creates one row for each column value. Row 1 of fig 2(a) splits into four rows each comprising value for one attribute. If there would have been three attributes, number of rows created in EAV model would have been three for each row of the relational model. The only variation is that in EAV no rows will be created for the attributes containing null values in relational model as EAV model only specifies the physical schema. EAV model is always supported by metadata which provides the logical schema for the EAV model. A relational model is easier to understand for the eyes of user. So, with the help of metadata (through pivoting [11]) an EAV modeled database can be transformed to relational model for the ease of user. EAV is capable of storing sparse and volatile data but it is not capable to handle relationships among class. There may be situation where one attribute is object of another class, in such cases EAV and metadata table cannot be stored as simple relational table but as object relation table. This enhancement done to EAV model is called EAV/CR model [12]. In EAV/CR model

Relationships among classes are maintained using either the concept of primary and foreign key or through the use of special table that contributes only for maintain relationships. The main weakness of the EAV/CR model is requirement for high data storage which can lead to many problems such as memory shortage and network congestion when applied on word wide level to achieve semantic interoperability.

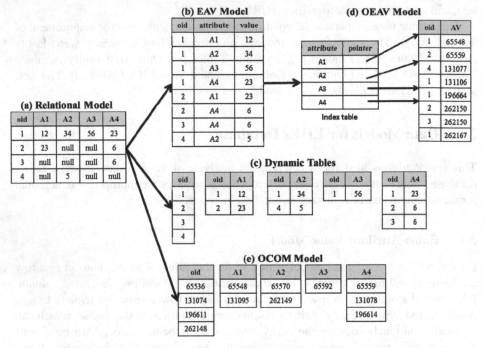

Fig. 2. Various data models considered for storing BIG DATA

Motive Behind EAV model. EAV model was proposed as an alternative for relation model to provide following functionalities.

No Sparseness. EAV model creates row for each non-null value present in relational model. Following this way no space is reserved for null values. Hence, there is no problem of sparseness.

Volatility. Adding new columns to the EAV model requires only creating a new row with attribute field containing the name of new attribute and value field will contain the corresponding value. For example, if we need to add another attribute named A5 with value 5 to the table then it can be done by just adding the name of new attribute in the attribute column of the EAV model. This way it is easy with volatility problem faced in relational model. So, EAV model can be used to store high dimensional data.

Cost Paid for EAV Model. After a rigorous survey, it has been found that EAV model lacks in some aspects from relation model which are as follow:

Difficult interpretation. As attributes of an entity splits into multiple equivalent number of rows (equivalent to non-null attributes), while retrieving the data user takes time to scan whole database for all rows in EAV model corresponding to a particular entity. So, it is difficult for user to understand the data contained in the EAV database.

Tools unavailability. No dedicated query tools are available in market to store and query EAV modeled databases. Tools available for relational model are used for EAV model too. So, querying an EAV modeled database is more complex as compared to relational model because queries available for relational model needs to be modified accordingly for querying EAV modeled database.

Heterogeneity. EAV model is capable of holding only one data type i.e. values contained in value column must be of same data type. So, different EAV database need to be constructed for different data types. This scenario is typically known as multi data type EAV schema.

Attribute & Entity centric Query. If we have queried for a particular attribute then whole database need to be searched for that attribute name in the attribute column. Similarly, in case of entity centric query whole database need to be searched for that entity name in the entity column. So, both types of queries take a lot of time to execute as whole database is searched always. This concludes that EAV is not search efficient.

Exponential growth for non-null values. If a relational model have three non-null attributes for an entity then EAV model will contain three rows (one for each attribute) containing three entries each (Entity, Attribute and Value). So, EAV model grows exponentially with the number of non-null values hence EAV requires more space.

3.2   Dynamic Tables

Dynamic tables is an extension of decomposition model [13-14] proposed by Copeland and Khoshafian. It works by dividing the n-ary relational model (i.e. containing n attributes) into n tables (as shown in fig 2 (c)). Among these n tables n-1 tables are composed of two columns; first for the primary key (oid) and second for a particular attribute (other than primary key). Other than these n-1 tables, there is one more table which contains only one column dedicated to store the contents of primary key of relational model. This table was an addition made to the decomposition model. The sole purpose of this table is to identify if a particular entity exists in the database or not.

EAV Issues Resolved by Dynamic Tables. Dynamic table overcome the problem of sparseness and volatility as EAV, but it is superior to EAV because of following advantages:

Attribute Centric Query. Whenever a query regarding a particular attribute is executed, dynamic tables responds very rapidly because it has to simple display the contents of one table which is dedicated for storing the particular attribute demanded.

Heterogeneity. Dynamic tables can handle heterogeneous data easily as it divides the relational table into many tables which can handle different data types.

Exponential growth for non-null values. Dynamic tables also expand with the expansion in non-null values but this expansion is less as compared to EAV.

Level of Interpretation. Understanding dynamic tables is easy than EAV because it stores one attribute in one table with primary key associated with it.

Author's Criticism for Dynamic Tables. Authors have identified that there are some limitations present in modeling data as dynamic tables.

Entity Centric Query. Dynamic tables have a positive point of storing the primary key in a separate dedicated table which will tell if a particular entity is present in the table or not. But if we found that the entity demanded exists in the database than we have to search all tables for the attribute values; no matter whether the entity have an entry in the attribute table or not (in case of null value).

Storage Requirement. Dynamic tables maintain only two entries per attribute value (i.e. primary key and attribute value) whereas EAV maintain three entries per attribute value (i.e. primary key, attribute name and attribute value). So, it seems like storage required for dynamic tables are less than EAV but this is not true. It has been proven experimentally by the authors in section 4 that dynamic tables consume more memory than EAV.

3.3   OEAV Model

Search inefficiency of EAV was a big issue being faced by the administrator. So, OEAV model was proposed by Paul et. al. in [9]. OEAV model is capable of dealing with the problems of sparseness and volatility (present in relational model) as EAV model but in a search efficient manner. The complete process of transforming EAV to OEAV is depicted diagrammatically in Appendix A. The conversion of EAV model to OEAV model as suggested by Paul et. al. starts with replacing attributes names in EAV model with the numeric codes stored in a table named Attribute Dictionary (defined by the administrator manually). The combinations of numeric attributes codes and corresponding value are joined as one value and all such values are stored under one column name, AV (Attribute Value) using the following steps.

1. Convert numeric attribute code and value to the equivalent binary representation comprising 16 bits each. For example (1, 11) is equivalent to (0000000000000001, 0000000000001011).
2. Concatenate both 16 bits codes to form a 32 bits code i.e. for above example it will be 00000000000000010000000000001011.
3. Now calculate the decimal equivalent of the 32 bits code i.e. for above example it will be 65547.

Finally, the transformed table is sorted based on column AV and then primary key (oid in fig 2(d)). An Index table is further created to point to first record (in the sorted table) of the corresponding attribute (in the index table). All the records corresponding to the particular attribute are found in the neighboring records, as some most significant digits of numeric codes will have same value for same attribute (initial 16 bits will be same for same attribute). So after sorting all neighboring elements will have same most significant digits until records for the next attribute is not started (having different most significant digits).

Functionalities Provided by OEAV Model. In addition to the functionalities (no sparseness and volatility) offered by EAV model, OEAV model offers following functionalities:

Attribute centric Query. If we have queried for a particular attribute then index table is searched for the particular attribute and correspond starting address for that attribute entries can be found. The sorted table is then searched till the most significant bits matches. So, searching time is quite less as it was in EAV model. So, OEAV model is search efficient.

Exponential growth for non-null values. This growth is less in OEAV model as compared to EAV model because in OEAV model only two entries (entity, AV) are created in each row whereas in   EAV model it is three (entity, attribute, value).

Author's Criticism for OEAV Model. In addition to the limitations (Difficult interpretation, Tools unavailability and Heterogeneity) in EAV model, authors identified that OEAV model has following limitations:

Calculation Overhead. For storing the values in AV column, numeric attribute codes and values need to be combined as one value by performing the calculation specified above. Also, while retrieving the data the value in AV column should be decoded to the original attribute code and value pair.

Inserting, Modification and Deletion. Insertion of a new record is quite complex in case of OEAV model since records are finally sorted based on the value of AV column. For the insertion of a new record, the database should be sorted again after the insertion of new record. Doing this requires shifting of all records for example if the new record for entity 5 is to be inserted in database shown in fig. 2(d) that contains non null attribute-value pair as (A1, 21), (A3, 7) and (A4, 22) then new table will look as shown in Fig 6 (new record entries are highlighted). For this insertion records need to be shifted down to accommodate new record entries (because list is sorted) and hence pointers in index table should also be updated to point to the first record of the corresponding attribute. Similar complex queries need to be executed for modification and deletion of records.

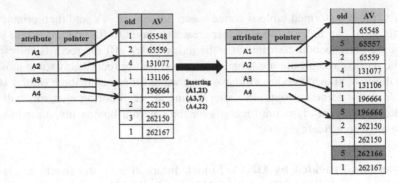

Fig. 3. Insertion of record in OEAV model

Entity Centric Query. For entity centric query, whole database need to be searched for all entries in sorted table corresponding to that entity. This process is same as it was in EAV model. So, there is no significant improvement in searching data related to particular entity.

Homogeneous data type. OEAV model is not capable of handling multiple data types. Moreover, it requires all values to be of numeric type so that sorting can be done easily.

Machine Dependency. As shown in fig 2(d) and fig 3, OEAV model construct an index table which contains pointer (address) to the first record of the corresponding attribute in the sorted table. If the database is reallocated to any other machine new address will be allocated to the OEAV database in new machine. So, for desired working all pointers values in the index table need to be reassigned the new address values. This makes the database machine dependent.

3.4   Optimized Column Oriented Model

OCOM emerged as a combination of positive points of decomposition model and OEAV model. The detailed process of transforming Relational model to OCOM is depicted diagrammatically in Appendix B. The model is constructed by decomposing the n-ary relational table (i.e. containing n columns) into 'n' attribute tables (as shown in fig 2(e)). The values contained in decomposed tables are then re-written as pairs of the position number of the attribute value in table and the corresponding attribute value for all non-null values. These pairs are then combined in one value by following the procedure used in OEAV model to convert attribute value pair in one value.

Highlights of OCOM Model. Other than no sparseness and volatility, OCOM model shows advantages of both OEAV model and Dynamic tables which are as follow:

Storage Requirement. OCOM model take minimum space for the storing all values as compared to all other models because it embeds the position value with the attribute value in one code (following the concatenate procedure of OEAV model). This statement is proved experimentally by the authors in section 4.

Exponential growth for non-null values. There is no exponential growth with non-null value as the value is coded and stored as a single value only. No other information (primary key or attribute name) is stored with that coded value; all information is contained in the coded value itself.

Machine Independent. OCOM involves no use of pointers, so the tables of OCOM are re-locatable.

Author's Criticism for OCOM Model. Authors have identified and proved experimentally that OCOM model reduces the memory consumed but at a cost of following:

Calculation Overhead. Similar to OEAV, OCOM also works with coded value. So, there is lot of calculations and thus the overhead.

Attribute Centric and Entity Centric Query. If we have queried for a particular attribute then it will accessed very quickly but for understanding the values of that attributes, all stored values must be decoded which is very complex. Performing an entity centric query is even more complex. Firstly we need to decode the table containing entities. After that all attribute values stored in all other decomposed tables need to be decoded. Only after following this complex process, details of particular entity can be found from all tables.

Homogeneous data type. OCOM require all values to be of type numeric which is not always possible. If any other data type needs to be stored, it has to be first coded to a numeric value. So, proper transformation tables need to be maintained for this. Moreover we will not be able to store the decimal value which is required in case of EHR. For example, body temperature is a decimal value.

Level of Interpretation. Understanding OCOM is very complex, even more complex than OEAV. In OEAV, we have pairs of primary key and coded value i.e. at least we can predict that the data is of which entity. But in case of OCOM, we have coded the value of primary key also and other attributes are coded according to the pair of position and attribute value.

4   Experimental Details

4.1   Data Preparation

A comparative analysis has been conducted for finding the suitability among various data models. Authors have performed an experiment on EHRs database containing 16 attributes where some attributes are of integer type, some contain string data and some are of Boolean data type. Complete details of the datset is shown in Table 1. The data is about the diabetic patients. The data collected is highly sparse, interrelated and hetrogeneous. This data has been collected from 5 local clinics and it contains 20,000 instances. Authors aim to increase the data size in future.

Table 1. Comparison of various data models for EHRs Database

S.No.	Attribute Name	Description	Data Type/ Domain
1	ID	Unique patient ID	Alpha numeric
2	Name	Name of the patient	String
3	Gender	Gender of the patient	Character (M-Male/ F-Female)
4	Age	Age of the patient	Numeric
5	Add	Address of the patient	Text
6	Weight	Weight of the patient	Numeric
7	Pregnancy	Is the patient pregnant?	Boolean
8	Thirst	Is Patient facing excessive thirst?	Character (E-Excessive, N-Normal)
9	Urination	Is Patient facing excessive Urination?	Character (E-Excessive, N-Normal)
10	BP	Blood Pressure of the patient	Numeric
11	FBS	Fasting Blood Sugar	Numeric
12	PBS	2 hour Postprandial Blood Sugar	Numeric
13	RBS	Random Blood Sugar	Numeric
14	OGT	Oral Glucose Tolerance	Numeric
15	HbA1c	Glycohemoglobin A1C	Numeric
16	Medication	Medication given to the patient	Text

The dataset was collected intially in relational model and then converted to EAV, OEAV, OCOM, dynamic table representations manually by the authors. In EAV, being one dedicated column for storing value, all attributes were stored as Text datatype. In OEAV and OCOM, a proper coding system were used to convert all attribute values into numeric For example, in case of gender M is replaced with 0 and F is replaced with 1. No datatype conversion is done in case of dynamic tables. Number of records in EAV and OEAV table are *16*20000 – Total number of null entries*. In case of OCOM and dynamic tables 16 different tables were constructed, each having number of records as *20000 – Total number of null entries of the dedicated attribute*.

4.2   Experiment Requirements

We have chosen MySQL Workbench 6.0 CE as a database management system for maintaining our database and executing queries on it. The operating system used is Windows 8.1. The processor of machine on which experiment is performed is AMD dual-core processor C60 with Turbo CORE technology up to 1.333 GHz. The RAM available in the machine is 2 GB DDR3 and hard disk is of 320 GB.

4.3   Modeling EHRs Database

Currently non-standardized EHRs schemas are adopted by most of the health organizations but ideally this creates a lot of problems. When some data needs to be communicated for the purpose of knowledge transfer, it will be meaningless until same terminology is followed by both organizations.

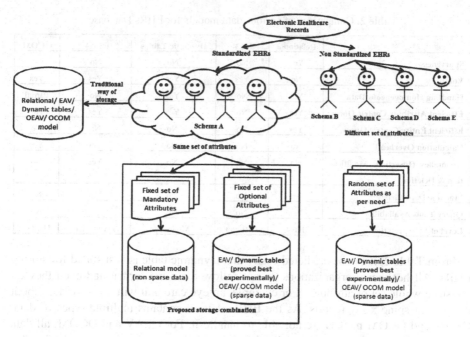

Fig. 4. Proposed Data Modeling for storing standardized and non-standardized EHRs

Organizations such as openEHR, CEN, ISO and HL7 [15-19] are working on this problem to provide a common standard schema for storing EHRs. To define a standard format, dual model approach comes into existence [5]. This approach was originally proposed by openEHR [15], later adopted by other standards such as CEN [16], ISO/EN 13606 [17-18] and HL7 [19]. In the dual model approach, data is defined using two different abstraction levels namely the Reference Model (RM) [5] and the Archetype Model (AM) [5]. In RM, the basic structures of various concepts used in medical terms are defined such as body mass index, blood pressure, and temperature. RM provides a generalized view of all the medical concepts which can be specialized in AM for individual use. AM defines various constraints that need to imposed on the structures defined by RM. Many archetypes can be designed based on same RM without making changes to the RM structure through AM. Archetypes also specify a defined set of mandatory and optional attributes.

After rigorous survey, various parameters (Sparseness, Volatility, Heterogeneity, Efficient Query, Overhead, Anomalies, Machine Dependency, Tools Availability and Interpretation) have been considered for comparing various data models for EHRs database (as shown in Table 2). Table 2 is prepared considering the discussion done about various models in section 3.An administrator can have a clear and quick review of merits and demerits of all models studied in this paper from Table 2.

Table 2. Comparison of various data models for EHRs Database

	Relational	EAV	Dynamic Tables	OEAV	OCOM
Sparseness	Yes	No	No	No	No
Volatility	No	Yes	Yes	Yes	Yes
Handling Homogeneous Data	Yes	No	Yes	No	Partial
Efficient Attribute Centric Query	Yes	No	Yes	Yes	No
Efficient Entity Centric Query	Yes	No	No	No	No
Calculation Overhead	No	No	No	Yes	Yes
Anomalies (Insertion, Modification & Deletion)	No	No	No	Yes	No
Machine Dependency	No	No	No	Yes	No
Query Tools Availability	Yes	No	No	No	No
Level of Interpretation	Easy	Difficult	Moderate	Very Difficult	Hardest

From Table 2, it can be predicted easily that Dynamic table is best suited for storing EHRs. All healthcare organizations do not follow the same format and even they do not store whole set of attributes. For example, an eye care unit will not save data about the patient spine x-ray reports. As the EHRs dataset contains multiple types of data, OEAV and OCOM models are not able to handle it. For OEAV and OCOM, all data other than numeric data (e.g. string, Boolean) is coded as numeric for example, "yes" is replaced by 1 and "no" is replaced by 0. Healthcare organizations mostly adopt a non-standardized schema for storing EHRs. In such cases, where non standardized EHRs are involved; dynamic tables shows the best results. For standardized EHRs, many researchers adopt EAV model [2] or a variant of EAV model but authors propose to adopt different model to store different type of attributes as shown in fig. 4. Mandatory attributes (no null values) can be stored in relational model. Dynamic model will be best suited to optional attributes (considering the comparison of Table 2).

Experiment in this research work is conducted on a non-standardized EHRs dataset because a major portion of our healthcare industries adopt non-standardized EHRs. As shown in fig. 4, authors propose to divide the storage of standardized EHRs in two types of models. Relational model being the best for non-sparse data has to be adopted for mandatory attributes. For non-sparse data, decision can be made based on the same experiment done in the next section on non-standardized EHRs (data stored in non-standardized dataset and optional attributes of standardized dataset is sparse by nature).

4.4   Performance Analysis

Modeling and managing BIG DATA effectively is measured on four parameters which are volume, velocity, variety and veracity, called 4 V's of BIG DATA [20]. Volume means that the size of data is very huge. Velocity signifies the analyzing speed of the data. Variety is about the different forms of data. Veracity denotes the correctness of data.

A data model can be judged based on two aspects i.e. volume and velocity. An efficient data model should be capable of handling large volume of data in a manner which can be analyzed very speedily. Variety is dependent on the type of source of data; it can be online (on machine) or offline (on papers). Data Models will be able to handle only data available electronically. Veracity defines the uncertainty present in data; data model is capable of storing data but the data may be entered incorrect by the user. The four models (EAV, Dynamic tables, OEAV and OCOM) discussed above for EHRs are compared on the volume and velocity parameters of BIG DATA (as shown in fig. 5).

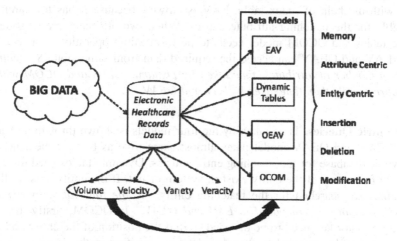

Fig. 5. Suitability of various data models considering EHRs data

The volume parameter is directly related with the memory acquired by the model and velocity is connected with how rapidly queries (attribute centric, entity centric, insertion, deletion and modification) can be executed on the model. As stated before, we have no dedicated query tool available for EAV, Dynamic Tables, OEAV, and OCOM models. So, we have to write relational query in such manner that when they are executed on EAV, Dynamic Tables, OEAV, and OCOM they give the desired output in the desired manner. For Attribute and Entity centric queries, selection and projection operation of relational model are used in an appropriate manner that suits various model. For insertion, deletion and modification of records in various models, DML (data manipulation language) queries were used in a suitable manner.

Memory Consumed. Storage space taken by each model is shown graphically in fig. 6(a). Relational model takes maximum space because dataset considered is highly sparse. OCOM takes least space as it stores the coded value in a single column (in multiple tables). Although dynamic tables follow the same approach of decomposing the table as OCOM but dynamic tables stores two columns per table which adds to the space consumed. An interesting comparison is seen in EAV and OEAV. At first look, it seems that OEAV will take less time than EAV because OEAV stores two entries for each non-null attribute value whereas EAV stores three entries per non-null attribute value. This is not true every time. OEAV constructs 32 bit code for each value regardless of space taken by attribute name and value (in case of EAV). This may lead to wastage of space.

Attribute Centric Queries. Comparison of the time taken to execute the attribute centric queries is shown through charts in fig. 6(b) and 6(c). When a single attribute is to be selected; Dynamic tables performed best as only one table was to be accessed. In OCOM also only one table is to be selected. OEAV can access the attribute records directly with the help of index table. EAV is slowest because it has to search the whole table for the matching attribute entries. When two attributes are considered, Dynamic tables and OCOM model needs to perform a join operation on two tables whereas EAV and OEAV can access the required data from same table. *So, with the increase in number of attributes, time taken by Dynamic tables and OCOM will increase more rapidly than time taken by EAV and OEAV.*

Entity Centric Queries. Time taken by the four models is shown through a chart in fig. 6(d). EAV and OEAV model takes almost same time as both of them need to search whole database for the matching entity entries. Dynamic tables need to search "oid" table first to see if the entity exist in system or not. If the entity exists, all dynamic tables are searched for the matching entity, so, *the time taken by dynamic tables will be more as compared to EAV and OEAV.* In OCOM, firstly the table containing primary key need to be decoded to find the position of the entity and then all tables need to be decoded to find the matching position in the (position, value) pair. This whole process is very complex and time consuming hence *OCOM will take the highest time for entity centric queries.*

Inserting New Record. Chart in fig 6(e) depicts the time taken by different models to insert a new record. Insertion in EAV model is simplest among all four models. To insert a record in EAV, triplets of entity, attribute and value are inserted simply. In case of dynamic tables a new record is inserted in "oid" table and records consisting of pair (primary key, value) are inserted in the dynamic tables for each non null attribute value. Hence dynamic tables take more time than EAV. OEAV and OCOM code the actual value of the attribute by following specified previously. So, OEAV and OCOM take more time than EAV and dynamic tables. For OEAV, in addition to insertion of coding value, sorting whole database and modification of index table is also required which add extra time to the insertion process. Thus, *inserting a record in OEAV is very complex and time consuming process.*

Updating Existing Record. Comparison of time taken by EAV, Dynamic tables, OEAV and OCOM model is shown in fig 6(f). To modify a record in EAV, it is firstly searched in the whole table and then updated. Dynamic tables require only specified attribute table to be searched for the required record and then it will be updated. OEAV model perform modification by firstly finding the required record (as coded value)

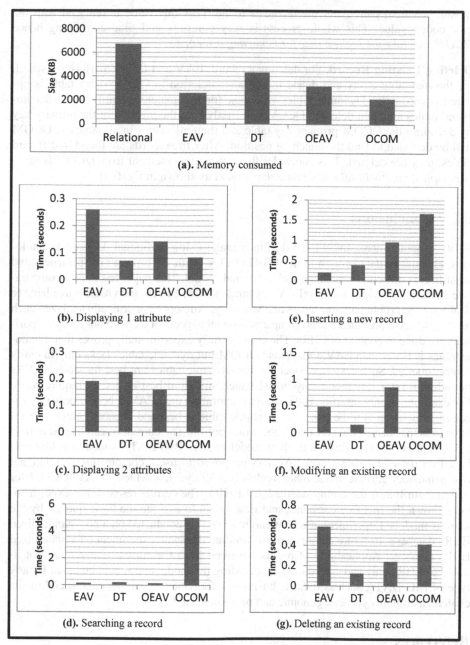

Fig. 6. Comparison of EAV, Dynamic Tables, OEAV and OCOM

and then it updates the record with new coded value. After this modification OEAV table needs to be sorted again. In OCOM, firstly the table containing primary key need to be decoded to find the position of the entity whose value is to be updated and then the dedicated attribute tables need to be decoded to find the matching position in the (position, value) pair. After the record is found, previous value is replaced with the new coded value. This whole process is very complex and time consuming hence *OCOM will take the highest time for modifying a record.*

Deleting Existing Record. For deleting a record, EAV and OEAV performs a search for the matching entity and delete it. But after deletion in OEAV index table is updated accordingly. Dynamic tables will look for the matching entity in all decomposed tables and then delete it. OCOM will firstly find the position of the primary key by decoding the OCOM primary key table and then all values in all tables of OCOM will be decoded to find the matching position. After the records are found in different tables; they are deleted. This concludes that deleting an element from *OCOM is highly complex, so, it will take more time than others* as shown in fig. 6(g).

5   Conclusions

EHRs generate increasingly large data to manage using traditional database technologies. Healthcare data being a form of BIG DATA (with the problem of sparseness and volatility) is difficult to maintain with relational model. This paper has done a comparative analysis of various models (EAV, Dynamic tables, OEAV and OCOM) available to deal with the problem of sparseness and volatility. Authors have critically analyzed each model and have provided a deep comparison on all aspects of a data model All comparisons are done in context of EHRs. The current study experimentally proves that dynamic tables is better than EAV, OEAV and OCOM but dynamic tables lacks only in entity centric queries. So far we can conclude that dynamic tables are best for storing EHRs but administrator can choose any model which suits him best as per his requirements. For example, if only insertion is the required operation then EAV is best, if only modification or deletion is to be done then dynamic table is best and if entity centric queries have to be performed then OCOM is the worst option. The current research aims to help a data administrator to find out the best model to store BIG DATA which suits him best based on the comparing criteria. Authors proposed to use a combination of relational (for mandatory attributes) and other models (EAV/ dynamic tables/ OEAV/ OCOM for optional attributes) for storing standardized EHRs. The current research has experimentally shown the memory consumed and time taken to execute the basic queries by the various models. Experiment in this research work was conducted on a non-standardized EHRs database. In future authors aim to simulate the same experiment on standardized EHRs database. Applicability of results retrieved from the current research are not limited to EHRs only and can be used for deciding the model that should be opted for any type of BIG DATA such as finance, business informatics, streaming data, social media content, astronomy surveys, genomic and proteomic studies.

References

1. Big Data, http://en.wikipedia.org/wiki/Big_data (accessed July 2014)
2. Batra, S., Sachdeva, S., Mehndiratta, P., Parashar, H.J.: Mining Standardized Semantic Interoperable Electronic Healthcare Records. In: Pham, T.D., Ichikawa, K., Oyama-Higa, M., Coomans, D., Jiang, X. (eds.) ACBIT 2013. CCIS, vol. 404, pp. 179–193. Springer, Heidelberg (2014)

3. Batra, S., Parashar, H.J., Sachdeva, S., Mehndiratta, P.: Applying data mining techniques to standardized electronic health records for decision support. In: Sixth International Conference on Contemporary Computing (IC3), pp. 510–515 (2013), doi:10.1109/IC3.2013.6612249
4. Xu, Y., Agrawal, R., Somani, A.: Storage and querying of e-commerce data. In: Proceedings of the 27th VLDB Conference, Roma, Italy (2001)
5. Beale, T., Heard, S.: The openEHR architecture: Architecture overview. In: The openEHR release 1.0.2, openEHR Foundation (2008)
6. Sachdeva, S., Bhalla, S.: Semantic Interoperability in Standardized Electronic Health Record Databases. ACM Journal of Data and Information Quality 3(1), Article 1 (2012)
7. Dinu, V., Nadkarni, P.: Guidelines for the effective use of entity-attribute-value modeling for biomedical databases. International Journal of Medical Informatics 76, 769–779 (2007)
8. Corwin, J., Silberschatz, A., Miller, P.L., Marenco, L.: Dynamic tables: An architecture for managing evolving, heterogeneous biomedical data in relational database management systems. Journal of the American Medical Informatics Association 14(1), 86–93 (2007)
9. Paul, R., et al.: Optimized Entity Attribute Value Model: A Search Efficient Representation of High Dimensional and Sparse Data. IBC 3(9), 1–6 (2009), doi:10.4051/ibc.2011.3.3.0009
10. Paul, R., Latiful Hoque, A.S.M.: Optimized column-oriented model: A storage and search efficient representation of medical data. In: Khuri, S., Lhotská, L., Pisanti, N. (eds.) ITBAM 2010. LNCS, vol. 6266, pp. 118–127. Springer, Heidelberg (2010)
11. Dinu, V., Nadkarni, P., Brandt, C.: Pivoting approaches for bulk extraction of Entity–Attribute–Value data. Computer Methods and Programs in Biomedicine 82, 38–43 (2006)
12. El-Sappagh, S.H., et al.: Electronic Health Record Data Model Optimized for Knowledge Discovery. International Journal of Computer Science 9(5), 329–338 (2012)
13. Copeland, G.P., Khoshafian, S.N.: A decomposition storage model. In: Proceedings of the 1985 ACM SIGMOD International Conference on Management of Data, pp. 268–279 (1985)
14. Khoshafian, S., Copeland, G., Jagodis, T., Boral, H., Valduriez, P.: A query processing strategy for the decomposed storage model. In: Proceedings of the Third International Conference on Data Engineering, pp. 636–643 (1987)
15. OpenEHR Community, http://www.openehr.org/ (accessed October 2013)
16. CEN - European Committee for Standardization: Standards, http://www.cen.eu/CEN/Sectors/TechnicalCommitteesWorkshops/CENTechnicalCommittees/Pages/Standards.aspx?param=6232&title=CEN/TC+251 (accessed July 2014)
17. ISO 13606-1.: Health informatics: Electronic health record communication. Part 1: RM (1st ed.) (2008)
18. ISO 13606-2.: Health informatics: Electronic health record communication. Part 2: Archetype interchange specification (1st ed.) (2008)
19. HL7. Health level 7, http://www.hl7.org (accessed October 2013)
20. http://www-01.ibm.com/software/data/bigdata/what-is-big-data.html (accessed July 2014)

Appendix A: Transforming EAV Model to OEAV Model

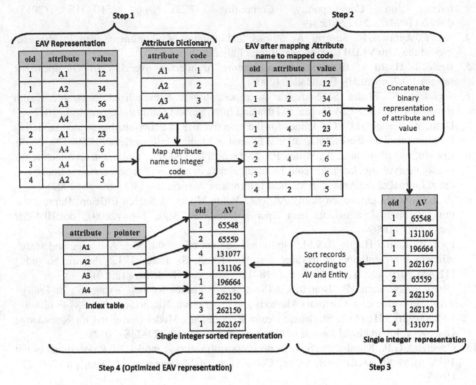

Appendix B: Transforming Relational Model to OCOM

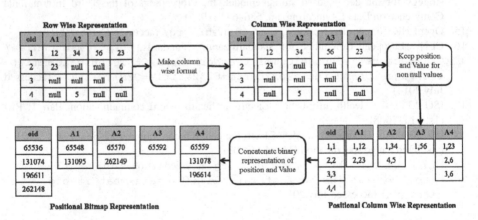

Sentiment Analysis of Twitter Feeds

Yogesh Garg and Niladri Chatterjee

Department of Mathematics,
Indian Institute of Technology, Delhi
New Delhi 110016

Abstract. This paper focuses on classifying tweets based on the sentiments expressed in them, with the aim to classify them into three categories: positive, negative and neutral. In particular, we investigate the relevance of using a two-step classifier and negation detection in the space of Twitter Sentiment analysis. An efficient sentiment analyzer is deemed to be a must in the era of big data where preponderance of electronic communication is a major bottleneck. Major difficulties in handling of tweets are, their limited size, and the cryptic style of writing that makes them difficult to comprehend at times. We have used different datasets publicly available online and designed a comprehensive set of pre-processing steps that make the tweets more amenable to Natural Language Processing techniques. Two classifiers are designed based on Naive-Bayes and Maximum Entropy classifiers, and their accuracies are compared on different feature sets. We feel that such classifiers will help business or corporate houses, political parties or analysts etc. to evaluate public sentiments about them and design appropriate policies to address their concerns.

1 Introduction

Sentiment Analysis refers to the use of text analysis and natural language processing to identify and extract subjective information in textual contents. There are two type of user-generated content available on electronic media (e.g. blogs, twitter) *facts* and *opinions*. By facts, we mean factual statements about certain topics, whereas opinions are user specific statement exhibiting positive or negative sentiments. Generally, while facts can be categorized using keywords, categorising opinions is harder. Various text analysis and machine learning techniques are used to mine statements from a document [1]. Sentiment Analysis finds its application in a variety of domains.

Business. Businesses may use sentiment analysis on blogs, review websites etc. to judge the market response of a product. This information may also be used for intelligent placement of advertisements. For example, if product "A" and "B" are competitors and an online merchant business "M" sells both, then "M" may advertise for "A" if the user displays positive sentiments towards "A", its brand or related products, or "B" if they display negative sentiments towards "A".

S. Srinivasa and S. Mehta (Eds.): BDA 2014, LNCS 8883, pp. 33–52, 2014.

Government. Governments and politicians can actively monitor public sentiments as a response to their current policies, speeches made during campaigns etc. This will help them make create better public awareness regarding policies and even drive campaigns intelligently.

Financial Markets. Public opinion regarding companies can be used to predict performance of their stocks in the financial markets. If people have a positive opinion about a product that a company "A" has launched, then the share prices of "A" are likely to go higher and vice versa. Public opinion can be used as an additional feature in existing models that try to predict market performances based on historical data.

In the present work we focus our analysis on sentiments expressed on Twitter. Twitter is an online social networking and micro-blogging service that enables users to create and read short messages, called "Tweets". It is a global forum with the presence of eminent personalities from the field of entertainment, industry and politics. People tweet about their life, events and express opinion about various topics using text messages limited to 140 characters. Registered users can read and post tweets, but any unregistered users can read them. Twitter can be accessed via Web, SMS, or mobile apps. Traditionally a large volume of research in sentiment analysis and opinion mining has been directed towards larger pieces of text like movie reviews. Sentiment Analysis in micro-blogging sphere is relatively new. From the perspective of Sentiment Analysis, we discuss a few characteristics of Twitter:

Length of a Tweet. The maximum length of a Twitter message is 140 characters. This means that we can practically consider a tweet to be a single sentence, void of complex grammatical constructs. This is a vast difference from traditional subjects of Sentiment Analysis, such as movie reviews.

Language used. Twitter is used via a variety of media including SMS and mobile phone apps. Because of this and the 140-character limit, language used in Tweets tend be more colloquial, and filled with slang and misspellings. Use of hashtags also gained popularity on Twitter and is a primary feature in any given tweet. Our analysis shows that there are approximately 1-2 hashtags per tweet, as shown in Tab. 3.

Data availability. Another difference is the magnitude of data available. With the Twitter API, it is easy to collect millions of tweets for training. There also exist a few datasets that have automatically and manually labelled the tweets [2], [3].

Domain of topics. People often post about their likes and dislikes on social media. These are not all concentrated around one topic. This makes twitter a unique place to model a generic classifier as opposed to domain specific classifiers that could be built using datasets such as movie reviews.

All of these characteristics of twitter make it different from other media, and unique to handle in comparison with other well known domains of text classification. At the same time, its huge applicability cannot be denied as it is one of the major communication platforms for collecting mass opinion. Consequently, in the era of big data, quick and efficient techniques for sentiment analysis appear to be a prime requirement. The present work is aimed at catering to this need.

This paper is organized as follows. In Section 2 we describe some of the previous works on sentiment analysis as carried out on texts from different domains. Section 3 describes the proposed methodology, including the pre- processing required, feature extraction, negation handling etc. for a dataset of 10000 tweets collected from different datasets. Section 4 presents the results or two classifiers, viz. Naive Bayes and Maximum Entropy classifiers. Section 5 concludes the paper with possible future directions.

2   Literature Review

Pang and Lee discuss in detail the current techniques in the area of sentiment analysis [1]. Much of it revolves around reviews for movies and products.

Go, Bhayani and Huang were among the first to explore sentiment analysis on Twitter [2]. They classify Tweets for a query term into negative or positive sentiment. They collect training dataset automatically from Twitter. To collect positive and negative tweets, they query twitter for happy and sad emoticons. Happy emoticons are different versions of smiling face, like ":)", ":-)", ":)", ":D", "=)" etc. Sad emoticons include frowns, like ":(", ":-(", ":(" etc. They try various features – unigrams, bigrams and Part-of-Speech and train their classifier on various machine learning algorithms – Naive Bayes, Maximum Entropy and Scalable Vector Machines and compare it against a baseline classifier by counting the number of positive and negative words from a publicly available corpus. They report that Bigrams alone and Part-of-Speech Tagging are not helpful and that Naive Bayes Classifier gives the best results.

Pak and Paroubek use a similar distant supervision technique to automatically collect the dataset from the web [4]. Apart from using happy and sad emoticons for positive and negative sentiment respectively, they also query tweets from accounts of 44 newspapers, like "New York Times", "Washington Posts" etc. to collect a training set of subjective tweets. They use unigrams and filtered n-grams for their classification. They also handle negations by attaching negative cues such as "no", "not" to the words preceding and following them. They report that both bigrams and negation handling help.

Koulompis, Wilson and Moore identify that use of informal and creative language make sentiment analysis of tweets a rather different task [5]. They leverage previous work done in hashtags and sentiment analysis to build their classifier. They use Edinburgh Twitter corpus[1] to find out most frequent hashtags. They manually classify these hashtags and use them to in turn classify the tweets. Apart from using n-grams and Part-of- Speech features, they also build a feature set from already existing MPQA[2] subjectivity lexicon and Internet Lingo

---

[1] http://demeter.inf.ed.ac.uk
[2] http://mpqa.cs.pitt.edu/lexicons/subj_lexicon/

Dictionary[3]. They report that the best results are seen with n-gram features with lexicon features, while using Part-of-Speech features causes a drop in accuracy.

Saif, He and Alani discuss a semantic based approach to identify the entity being discussed in a tweet, like a person, organization etc. [6] They also demonstrate that removal of stop words is not a necessary step and may have undesirable effect on the classifier.

All of the aforementioned techniques rely on n-gram features. It is unclear that the use of Part-of-Speech tagging is useful or not. To improve accuracy, some employ different methods of feature selection or leveraging knowledge about micro-blogging. In contrast, we improve our results by using more basic techniques used in Sentiment Analysis, like stemming, two-step classification and negation detection and scope of negation.

Negation detection is a technique that has often been studied in sentiment analysis. Negation words like "not", "never", "no" etc. can drastically change the meaning of a sentence and hence the sentiment expressed in them. Due to presence of such words, the meaning of nearby words becomes opposite. Such words are said to be in the scope of negation. Many researches have worked on detecting the scope of negation.

The scope of negation of a cue can be taken from that word to the next following punctuation. Councill, McDonald and Velikovich discuss a technique to identify negation cues and their scope in a sentence [7]. They identify explicit negation cues in the text and for each word in the scope. Then they find its distance from the nearest negative cue on the left and right.

3   Methodology

We use different feature sets and machine learning classifiers to determine the best combination for sentiment analysis of twitter. We also experiment with various pre-processing steps like – punctuations, emoticons, twitter specific terms and stemming. We investigated the following features – unigrams, bigrams, trigrams and negation detection. We finally train our classifier using various machine-learning algorithms – Naive Bayes and Maximum Entropy.

We use a modularized approach with feature extractor and classification algorithm as two independent components. This enables us to experiment with different options for each component. Different steps taken in the entire process are illustrated in Fig. 1.

3.1   Datasets

One of the major challenges in Sentiment Analysis of Twitter is to collect a labelled dataset. Researchers have made public the following datasets for training and testing classifiers.

---

[3] http://idahocsn.org/wp-content/uploads/2014/03/InternetLingoDictionary.pdf

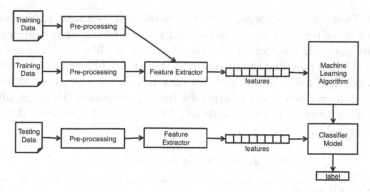

Fig. 1. Schematic Diagram of Methodology

Twitter Sentiment Corpus. This is a collection of 5513 tweets collected for four different topics, namely, Apple, Google, Microsoft, Twitter It is collected and hand-classified by Sanders Analytics LLC [3]. Each entry in the corpus contains, Tweet id, Topic and a Sentiment label. We use Twitter-Python library[4] to enrich this data by downloading data like Tweet text, Creation Date, Creator etc. for every Tweet id. Each Tweet is hand classified by an American male into the following four categories. Tweets displaying negative or positive sentiments are labelled accordingly. If there is no sentiment displayed, the tweet is marked neutral. The tweets that do not talk about the topic it was quried for or are not in English are manually labelled as irrelevant. For the purpose of our experiments, we consider Irrelevant and Neutral to be the same class. Illustration of Tweets in this corpus is show in Tab. 1.

Positive. For showing positive sentiment towards the topic
Positive. For showing no or mixed or weak sentiments towards the topic
Negative. For showing negative sentiment towards the topic
Irrelevant. For non English text or off-topic comments

Table 1. Overview of Twitter Sentiment Corpus

Class	Count	Example
neg	529	#Skype often crashing: #microsoft, what are you doing?
neu	3770	How #Google Ventures Chooses Which Startups Get Its $200 Million http://t.co/FCWXoUd8 via @mashbusiness @mashable
pos	483	Now all @Apple has to do is get swype on the iphone and it will be crack. Iphone that is

---

[4] https://github.com/bear/python-twitter – This library provides a pure Python interface for the Twitter API.

Stanford Twitter. This corpus of tweets, developed by Sanfords Natural Language processing research group, is publically available [2]. The training set is collected by querying Twitter API for happy emoticons like ":)" and sad emoticons like ": (" and labelling them positive or negative. The emoticons were then stripped and Re-Tweets and duplicates removed. It also contains around 500 tweets manually collected and labelled for testing purposes. We randomly sample and use 5000 tweets from this dataset. An example of Tweets in this corpus are shown in Tab. 2.

Table 2. Overview of Stanford Corpus

Class	Count	Example
neg	2501	Playing after the others thanks to TV scheduling may well allow us to know what's go on, but it makes things look bad on Saturday nights
pos	2499	@francescazurlo HAHA!!! how long have you been singing that song now? It has to be at least a day. i think you're wildly entertaining!

3.2 Pre-Processing

Tweets are generally made using colloquial language and use of emoticons, hashtags, misspellings etc. is very frequent. such a form of text is not readily usable for machine learning purposes. It becomes important to normalize the text by applying a series of pre-processing steps. We have applied an extensive set of pre-processing steps to decrease the size of the feature set to make it suitable for learning algorithms. We have given an illustration of a tweet in Fig. 2. The frequency of such patterns per tweet, cut by datasets is given in Tab. 3. Following which we also give a brief description of pre-processing steps taken.

Table 3. Frequency of Features per Tweet

Features	Twitter Sentiment Avg.	Max.	Stanford Corpus Avg.	Max.	Both Avg.	Max.
Handles	0.6761	8	0.4888	10	0.5804	10
Hashtags	2.0276	13	0.0282	11	1.0056	13
Urls	0.4431	4	0.0452	2	0.2397	4
Emoticons	0.0550	3	0.0154	4	0.0348	4
Words	14.4084	31	13.2056	33	13.7936	33

Hashtags. A hashtag is a word or an un-spaced phrase prefixed with the hash symbol (#). These are used to both naming subjects and phrases that are currently in trending topics e.g. #iPad, #news. Regular expressions can be used to find such patterns. #(\w+) was used to find hashtags and they were replaced by the following expression: HASH_\1

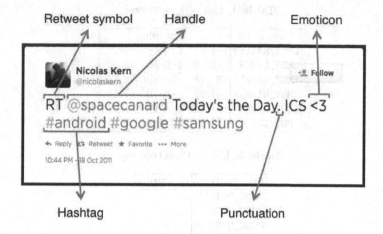

Fig. 2. Illustration of a Tweet with various features

Handles. Every Twitter user has a unique username. Any thing directed towards that user can be indicated be writing their username preceded by @. Thus, these are like proper nouns e.g. @Apple. The regular expression used to find out handles is @(\w+) and the replaced expression is HNDL_\1

URLs. Users often share hyperlinks in their tweets. Twitter shortens them using its in-house URL shortening service, like http://t.co/FCWXoUd8 – such links also enables Twitter to alert users if the link leads out of its domain. From the point of view of text classification, a particular URL is not important. However, presence of a URL can be an important feature. Regular expression for detecting a URL is fairly complex because of different types of URLs that can be there, but because of Twitters shortening service, we can use a relatively simple regular expression, (http|https|ftp)://[a-zA-Z0-9\./]+ which we replace by a simple word, URL.

Emoticons. Use of emoticons is very prevalent throughout the web, more so on micro-blogging sites. We identify the following emoticons and replace them with a single word. A list of emoticons we are currently detecting is given in Tab. 4. All other emoticons would be ignored.

Punctuations. Although not all Punctuations are important from the point of view of classification but some of these, like question mark, exclamation mark can also provide information about the sentiments of the text. We replace every word boundary by a list of relevant punctuations present at that point. A list of punctuations currently identified is given in Tab. 5. We also remove any single quotes that might exist in the text.

Table 4. List of Emoticons

Emoticons	Examples
EMOT_SMILEY	:-) :) (: (-:
EMOT_LAUGH	:-D :D X-D XD xD
EMOT_LOVE	<3 :*
EMOT_WINK	;-) ;) ;-D ;D (; (-;
EMOT_FROWN	:-(:((: (-:
EMOT_CRY	:,(:'(:"(:((

Table 5. List of Punctuations

Punctuations	Examples
PUNC_DOT	.
PUNC_EXCL	!
PUNC_QUES	?
PUNC_ELLP	...

Repeating Characters. People often use repeating characters while using colloquial language, like "Im in a hurrryyyyy", "We won, yaaayyyyy!" As our final pre-processing step, we replace characters repeating more than twice as two characters.

Regular Expression: (.)\1{1,}
Replace Expression: \1\1

Reduction in Feature Space. It is important to note that by applying these pre-processing steps, we are reducing our feature set otherwise it can be too sparse. The decrease in size of the feature set due to each step is given in Tab. 6. The final row concludes the total reduction in feature set if we apply all the pre-processing steps.

Table 6. Number of words before and after pre-processing

Pre-processing	Twitter Sentiment		Stanford Corpus		Both	
	Words	Reduction	Words	Reduction	Words	Reduction
None	19128		15910		31832	
Hashtags	18649	2.50%	15550	2.26%	31223	1.91%
Handles	17118	10.51%	13245	16.75%	27383	13.98%
Urls	16723	12.57%	15335	3.61%	29083	8.64%
Emoticons	18631	2.60%	15541	2.32%	31197	1.99%
Punctuations	13724	28.25%	11225	29.45%	22095	30.59%
Repeatings	18540	3.07%	15276	3.98%	30818	3.19%
All	11108	41.93%	8646	45.66%	16981	46.65%

3.3 Stemming Algorithms

All stemming algorithms are of the following major types affix removing, statistical and mixed. The first kind, Affix removal stemmer, is the most basic one. These apply a set of transformation rules to each word in an attempt to cut off commonly known prefixes and / or suffixes. A trivial stemming algorithm would be to truncate words at nth symbol. But this obviously is not well suited for practical purposes. J.B. Lovins described first stemming algorithm in 1968 [8]. It defines 294 endings, each linked to one of 29 conditions, plus 35 transformation rules . For the word being stemmed, an ending with a satisfying condition is found and removed. Another famous stemmer used extensively is described in the next section.

Porter Stemmer. Another well-known stemming algorithm is Martin Porter's algorithm [9]. It was published in July 1980 and still remains the de facto algorithm for English stemming. It offers excellent trade-off between speed, readability, and accuracy. It uses a set of around 60 rules applied in 6 successive steps. An important feature to note is that it doesnt involve recursion. The steps in the algorithm are described in Tab. 7.

Table 7. Porter Stemmer Steps

1.	Gets rid of plurals and -ed or -ing suffixes
2.	Turns terminal y to i when there is another vowel in the stem
3.	Maps double suffixes to single ones: -ization, -ational, etc.
4.	Deals with suffixes, -full, -ness etc.
5.	Takes off -ant, -ence, etc.
6.	Removes a final e

Lemmatization. Lemmatization is the process of normalizing a word rather than just finding its stem. In the process, a suffix may not only be removed, but may also be substituted with a different one. It may also involve first determining the part-of-speech for a word and then applying normalization rules. It might also involve dictionary look-up. For example, verb saw would be lemmatized to see and the noun saw will remain saw. For our purpose of classifying text, stemming should suffice.

3.4 Features

A wide variety of features can be used to build a classifier for tweets. The most widely used and basic feature set is word n-grams. However, there is a lot of domain specific information present in tweets that can also be used for classifying them. We have experimented with two sets of features:

Unigrams. Unigrams are the simplest features that can be used for text classification. A Tweet can be represented by a multiset of words present in it. We have used the presence of a word itself, not how many times it occurred, as feature. Previous researchers have found that presence of unigrams yields better results than repetition [1]. This also helps us to avoid having to scale the data, which can considerably decrease training time [2]. The cumulative distribution of words in our dataset is illustrated in Fig. 3.

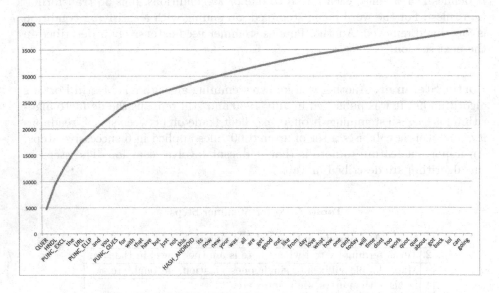

Fig. 3. Cumulative Frequency Plot for 50 Most Frequent Unigrams

We also observe that the unigrams nicely follow Zipfs law. It states that in a corpus of natural language, the frequency of any word is inversely proportional to its rank in the frequency table. A plot of log frequency versus log rank of the words appearing in our dataset is shown in Fig. 4. We notice that a linear trendline fits well with the data, which leads us to verify Zipf's law. The fitted line is given by the following equation:

$$log(f) = -0.9799log(r) + 3.9838 \tag{1}$$

where f is the frequency a word and r is the rank of the word in the frequency table.

$$f = 10^{3.9838}r^{-0.9799} \tag{2}$$

from which we can say that

$$f \propto \frac{1}{r} \tag{3}$$

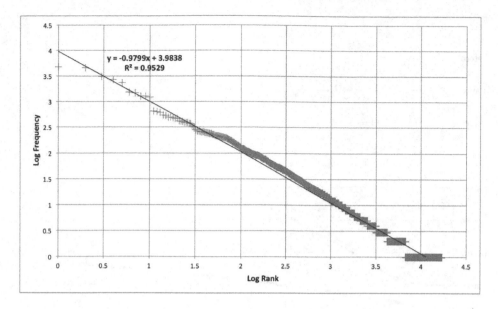

Fig. 4. Zipf's Law - Log Frequency versus Log Rank plot for unigrams

N-grams. N-gram refers to an n-long sequence of words. Probabilistic Language Models based on Unigrams, Bigrams and Trigrams can be successfully used to predict the next word given a current context of words. In the domain of sentiment analysis, the performance of N-grams is unclear. According to Pang et al., some researchers report that unigrams alone are better than bigrams for classification movie reviews, while some others report that bigrams and trigrams yield better product-review polarity classification [1].

As the order of the n-grams increases, they tend to be more and more sparse. Based on our experiments, we find that number of bigrams and trigrams increase much more rapidly than the number of unigrams with the number of Tweets. Fig. 5 shows the number of n-grams versus number of Tweets. We can observe that bigrams and trigrams increase almost linearly where as unigrams are increasing logarithmically.

Because higher order n-grams are sparsely populated, we decide to trim off the n-grams that are not seen more than once in the training corpus, because chances are that these n-grams are not good indicators of sentiments. After the filtering out non-repeating n-grams, we see that the number of n-grams is considerably decreased and equals the order of unigrams, as shown in Fig. 6.

50 most frequent bigrams and trigrams in our dataset with their respective cumulative frequency distributions are shown in Fig. 7 and Fig. 8. Relevant n-grams like "the,new", "thank,you", "HNDL,thanks", "HNDL,thanks,EXCL", "HNDL,get,the" can be seen in these lists.

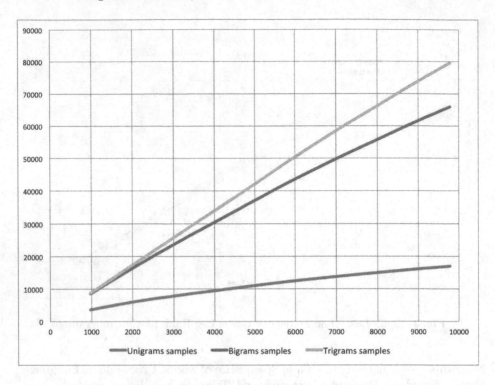

Fig. 5. Number of n-grams vs. Number of Tweets

Negation Handling. The need of negation detection in sentiment analysis can be illustrated by the difference in the meaning of the phrases, "This is good" vs. "This is not good" Handling the negation consists of two tasks Detection of explicit negation cues and the scope of negation of these words.

Councill et al. look at whether negation detection is useful for sentiment analysis and also to what extent is it possible to determine the exact scope of a negation in the text [7]. They describe a method for negation detection based on Left and Right Distances of a token to the nearest explicit negation cue.

Detection of Explicit Negation Cues. To detect explicit negation cues, we are looking for the following words in Tab. 8. The search is done using regular expressions.

Scope of Negation. Words immediately preceding and following the negation cues are the most negative and the words that come farther away do not lie in the scope of negation of such cues. We define left and right negativity of a word as the chances that meaning of that word is actually the opposite. Left negativity depends on the closest negation cue on the left and similarly for Right negativity. Fig. 9 illustrates the left and right negativity of words in a tweet.

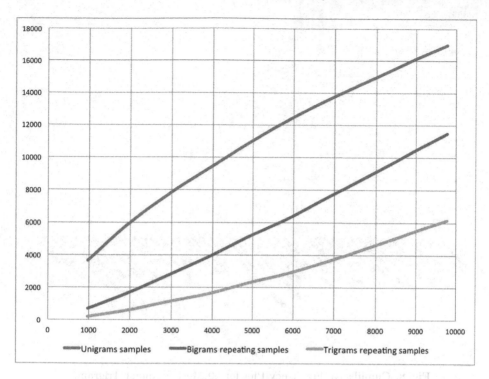

Fig. 6. Number of repeating n-grams vs. Number of Tweets

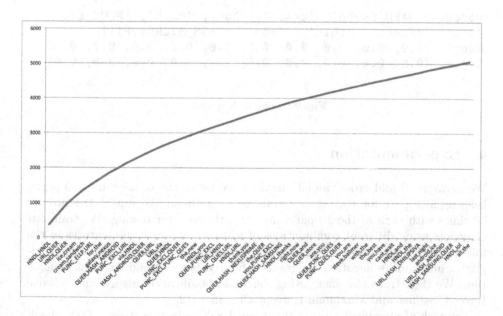

Fig. 7. Cumulative Frequency Plot for 50 Most Frequent Bigrams

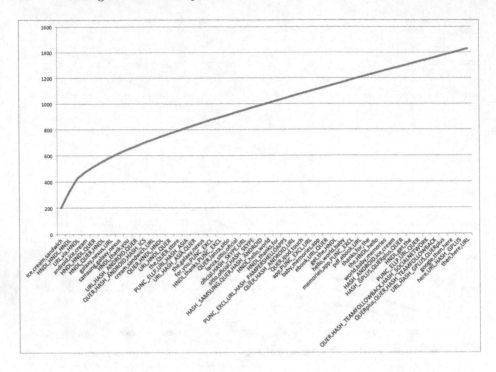

Fig. 8. Cumulative Frequency Plot for 50 Most Frequent Trigrams

```
Words: ['HASH_Skype', 'crash', 'too', 'much', 'PUNC_EXCL',
'not', 'expect', 'this', 'from', 'HASH_MICROSOFT']
Neg_l: [0.0, 0.0, 0.0, 0.0, 0.0, 1.0, 0.9, 0.8, 0.7, 0.6]
Neg_r: [0.5, 0.6, 0.7, 0.8, 0.9, 1.0, 0.0, 0.0, 0.0, 0.0]
```

Fig. 9. Scope of Negation

4   Experimentation

We perform 10-fold cross validation where we break the dataset into 10 parts, keep aside 1 part for testing and train on 9 of them. We repeat the process 10 times with each of the 10 parts used exactly once for testing. We train our classifiers using different combinations of features. We take the features in the following combinations – only unigrams, unigrams + filtered bigrams and trigrams, unigrams + negation, unigrams + filtered bigrams and trigrams + negation. We then train classifiers using different classification algorithms – Naive Bayes Classifier and Maximum Entropy Classifier [12].

The task of classification of a tweet can be done in two steps – first, classifying "neutral" (or "subjective") vs. "objective" tweets and second, classifying objective tweets into "positive" vs. "negative" tweets. We also trained 2 step

Table 8. Explicit Negation Cues

S.No.	Negation Cues
1.	never
2.	no
3.	nothing
4.	nowhere
5.	noone
6.	none
7.	not
8.	havent
9.	hasnt
10.	hadnt
11.	cant
12.	couldnt
13.	shouldnt
14.	wont
15.	wouldnt
16.	dont
17.	doesnt
18.	didnt
19.	isnt
20.	arent
21.	aint
22.	Anything ending with "n't"

classifiers. The accuracies for each of these configurations are shown in Fig. 10, we discuss these in detail below.

4.1 Naive Bayes

Naive Bayes classifier is the simplest and the fastest classifier. Many researchers [2], [4] claim to have achieved best results using this classifier.

For a given tweet, if we need to find the label for it, we find the probabilities of all the labels, given that feature and then select the label with maximum probability.

$$label_{NB} := argmax_{label} P(label|features) \qquad (4)$$

In order to find the probability for a label, this algorithm first uses the Bayes rule to express $P(label|features)$ in terms of $P(label)$ and $P(features|label)$ as,

$$P(label|features) = \frac{P(label) * P(features|label)}{P(features)} \qquad (5)$$

Making the 'naive' assumption that all the features are independent,

$$P(label|features) = \frac{P(label) * P(f_1|label) * ... * P(f_n|label)}{P(features)} \qquad (6)$$

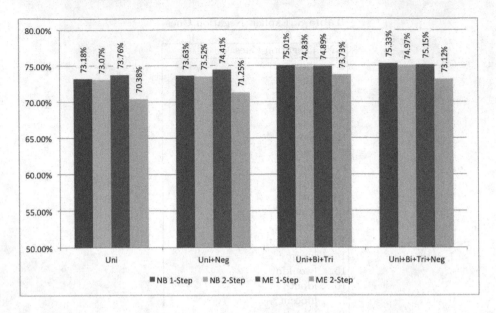

Fig. 10. Accuracy for Naive Bayes and Maximum Entropy Classifier

Rather than computing P(featues) explicitly, we can just calculate the numerator for each label, and normalize them so they sum to one:

$$P(label|features) = \frac{P(label) * P(f_1|label) * ... * P(f_n|label)}{\sum_l (P(l) * P(f_1|l) * ... * P(f_n|l))} \quad (7)$$

The results from training the Naive Bayes classifier are shown below in Fig. 10. The accuracy of Unigrams is the lowest at 73.18%. The accuracy increases if we also use Negation detection (73.63%) or higher order n-grams (75.01%). We see that if we use both Negation detection and higher order n-grams, the accuracy is increases again (75.33%). We can also note that accuracies for 2-step classifier are marginally lesser than those for corresponding 1-step.

We have also shown Precision vs. Recall values for Naive Bayes classifier corresponding to different classes Negative, Neutral and Positive in Fig. 11. The solid markers show the precision vs. recall for single step classifier and hollow markers show the affect of using double step classifier. Different points are for different feature sets. We can see that Precision for "Neutral" Class has come down and the same for "Negative" and "Positive" has gone up. Basically, the difference in Precision of "Neutral" and "Objective" has reduced, with a slight decrease in their respective Recall.

4.2   Maximum Entropy Classifier

This classifier works by finding a probability distribution that maximizes the likelihood of testable data. This probability function is parameterized by weight

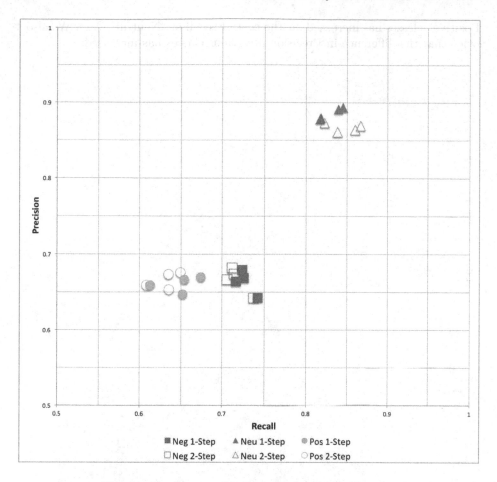

Fig. 11. Precision vs. Recall for Naive Bayes Classifier

vector. The optimal value of which can be found out using the method of Lagrange multipliers.

$$P(label|features) = \frac{\sum_i w_i f_i(label)}{\sum_{l \in labels} \sum_i w_i f_i(l)} \qquad (8)$$

The results from training the Naive Bayes classifier are shown below in Fig. 10. Accuracies follow a similar trend as compared to Naive Bayes classifier. Unigram is the lowest at 73.76% and we see an increase for negation detection at 74.41%. The maximum is achieved with unigrams, filtered bigrams and filtered trigrams at 74.89% closely followed by n-grams and negation at 75.15%. Once again, the accuracies for double step classifiers are considerably lower.

Precision versus Recall map is also shown for maximum entropy classifier in Fig. 12. Here we see that by using 2-Step classifier, the Recall of "Negative" and

"Positive" classes has increased, at the cost of recall of "Neutral" class. We also notice that the difference in Precision of various classes has increased.

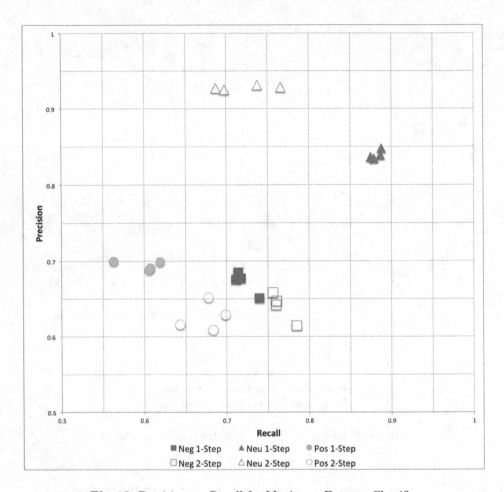

Fig. 12. Precision vs. Recall for Maximum Entropy Classifier

5   Conclusion

In this paper, we created a sentiment classifier for twitter using labelled data sets. We also investigate the relevance of using a double step classifier and negation detection for the purpose of sentiment analysis.

Our baseline classifier that uses just the unigrams achieves an accuracy of around 73.00%. Accuracy of the classifier increases if we use negation detection or introduce bigrams and trigrams. Thus we can conclude that both Negation Detection and higher order n-grams are useful for the purpose of text classification. Moreover, if we use both n-grams and negation detection, the accuracy is higher than both of them individualy. We also note that Single step classifiers out perform double step classifiers. Maximum Entropy Classifier poerforms better than Naive Bayes Classifier in case of unigrams and negation detection, but if we use higher n-grams also, Naive Bayes Classifier is seen to perform better than Maximum Entropy Classifier. We achieve the best accuracy of 75.33% in the case of Unigrams + Bigrams + Trigrams + Negation, trained on Naive Bayes Classifier.

The amount of data available on twitter that we can use for classification purposes is huge. It needs to be seen how our method works on big data. Appropriate distributed computing algorithms need to be developed for the implementation of the algorithms. This is one aspect that we are examining currently. Apart from this, following is a list of ideas we would like to explore in the future.

Investigating Support Vector Machines. Several papers have discussed the results using Support Vector Machines (SVMs) also. The next step would be to test our approach on SVMs. However, Go, Bhayani and Huang have reported that SVMs do not increase the accuracy [2].

Building a classifier for Hindi tweets. There are many users on Twitter that use primarily Hindi language. The approach discussed here can be used to create a Hindi language sentiment classifier.

Improving Results using Semantics Analysis. Understanding the role of the nouns being talked about can help us better classify a given tweet. For example, "Skype often crashing: microsoft, what are you doing?" Here Skype is a product and Microsoft is a company. We can use semantic labellers to achieve this. Such an approach is discussed by Saif, He and Alani [6].

References

1. Pang, B., Lee, L.: Opinion mining and sentiment analysis. Foundations and Trends in Information Retrieval 2(1-2), 1–135 (2008)
2. Go, A., Bhayani, R., Huang, L.: Twitter sentiment classification using distant supervision. Processing, 1–6 (2009)
3. Sanders, N.: Twitter sentiment corpus. Sanders Analytics, http://www.sananalytics.com/lab/twitter-sentiment/
4. Pak, A., Paroubek, P.: Twitter as a corpus for sentiment analysis and opinion mining, vol. 2010, pp. 1320–1326 (2010)
5. Kouloumpis, E., Wilson, T., Moore, J.: Twitter sentiment analysis: The good the bad and the omg! In: ICWSM, vol. 11, pp. 538–541 (2011)
6. Saif, H., He, Y., Alani, H.: Semantic sentiment analysis of twitter. In: Cudré-Mauroux, P., Heflin, J., Sirin, E., Tudorache, T., Euzenat, J., Hauswirth, M., Parreira, J.X., Hendler, J., Schreiber, G., Bernstein, A., Blomqvist, E. (eds.) ISWC 2012, Part I. LNCS, vol. 7649, pp. 508–524. Springer, Heidelberg (2012)
7. Councill, I.G., McDonald, R., Velikovich, L.: What's great and what's not: learning to classify the scope of negation for improved sentiment analysis. In: Proceedings of the Workshop on Negation and Speculation in Natural Language Processing, pp. 51–59. Association for Computational Linguistics (2010)
8. Smirnov, I.: Overview of stemming algorithms. In: Mechanical Translation (2008)

9. Porter, M.F.: An algorithm for suffix stripping. Program: electronic library and information systems 40(3), 211–218 (2006)
10. Gokulakrishnan, B., Priyanthan, P., Ragavan, T., Prasath, N., Perera, A.: Opinion mining and sentiment analysis on a twitter data stream. In: IEEE 2012 International Conference on Advances in ICT for Emerging Regions, ICTer (2012)
11. Quinlan, J.R.: C4. 5: programs for machine learning, vol. 1. Morgan kaufmann (1993)
12. Bird, S., Klein, E., Loper, E.: Natural language processing with Python. O'Reilly Media, Inc. (2009)

Graphical Analysis and Visualization of Big Data in Business Domains

Divanshu Gupta, Avinash Sharma, Narayanan Unny, and Geetha Manjunath

Xerox Research Center India
Bangalore, India
firstname.lastname@xerox.com

Abstract. Most efforts towards analyzing Big Data assume data parallel applications and handle the large volumes of data using Hadoop–like systems. However, Big Data is actually characterized by the 4V's – Volume, Variety, Velocity and Veracity. We propose a Big Data Stack and analytics solution that particularly caters to this important problem of addressing Variety and Velocity aspects of data by exploiting inherent relationship among data elements. A unique approach that we propose to take is to integrate and model the data using non-planar graphs and discover new insights through sophisticated graph analytics techniques. We have integrated the stack with an intuitive visualization toolkit that enables focused exploration of data, through query and selective visualization - which will be demonstrated.

1 Introduction

Efficient Information Management is one of the compelling problems of enterprises even today. When a company grows through a merger/acquisition or decides to modernize its IT portfolio by leveraging new technologies, it is imperative that they assimilate the older/legacy applications and data sources to the newer infrastructure. Use of such related but independent applications results in diverse data formats and schema. All the above reasons create a lot of heterogeneity in enterprise data sources. This hampers the speed of making IT/business decisions owing to the inability to see a single version of truth across the multiple data sources.

Data Integration from diverse data sources remains to be an important technical problem despite several earlier efforts. It has been a long standing challenge for the database community and has now gained further prominence with the growth of unstructured information from the Internet (blogs, portals, emails), diverse media types (video, audio, images) and new mechanisms to information (RSS,REST). The term "Big Data" is now a popular way to refer to massive digital information available in both structured and unstructured form integrated from multiple, diverse, dynamic sources of information. The applicability of Big Data applications is almost in all spheres - business, government, consumer – and spans application areas such as Web Mining, Social Media Analytics, Mobile Data Personalization, Customer Care, Healthcare, eGovernance, Datacenter Management, etc.

S. Srinivasa and S. Mehta (Eds.): BDA 2014, LNCS 8883, pp. 53–56, 2014.

Most efforts towards analyzing Big Data assume data parallel applications and handle the large volumes of data using Hadoop–like systems. However, actual data challenges faced by business problems is in deriving insights from fusing different variety of sources of time-varying data. We propose a Big Data Stack and analytics solution that particularly caters to this important problem of addressing Variety and Velocity aspects of data by exploiting inherent relationship among data elements. The proposed stack enables both goal driven data exploration as well as nonobvious insight discovery.

2   Solution Overview

Fig. 1 gives a high level view of the proposed stack. The GaBiD (Graphical Analysis of Big Data) stack is mainly targeted at data fusion and analysis of data originating from (a) structured data sources with diverse data schema (b) semi structured and (c) unstructured data types like text, image, video. A unique approach that we propose is to integrate and model the data using non-planar graphs and discover new insights through sophisticated graph analytics techniques.

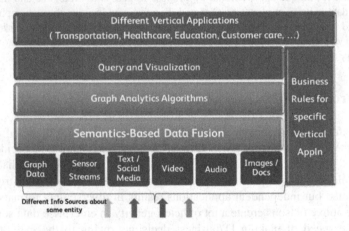

Fig. 1. Proposed GaBiD stack for data analysis

While some business problems inherently exhibit graphical structure (social network, bus routes, etc.), other business problems (such as creation of an entity profile using different data sources, or understanding business processes) can be solved using an induced graph using semantic-data fusion.(RDF [1]) to capture basic data facts (from the data sources) and knowledge of the domain (from domain experts). Use of a semantic data model helps capture the semantic heterogeneity and differences in the logical representation of data. Each data type can be injected into the bottom layer of the stack with appropriate one-time transformation to semantic models (by configuring different adaptors for different data types) and this results in a semantic graph representation of the fused data. This fused graph is passed through graph analytics to discover more insights. Graphical representation of data enables intuitive visualization and goal-oriented exploration of data through progressive queries.

We have integrated the stack with an intuitive visualization toolkit   that supports multiple graph clustering algorithms and enables focused exploration of data, through query and selective visualization. This stack and visualization tool will be demonstrated using data on customer care transactions.

3   Related Work

Semantic Data Integration has been a topic of research in database communities for 2 decades now. Use of RDF for Data Integration has also been tried in some experimental setups such as [3,4]. [5] proposes use of RDF based integration for HealthCare Systems. The D2R system [6] is an open source tool that provides a virtual RDF model over relational databases – enabling use of existing relational databases as RDF stores. Ontograte [7] provides a framework using which human experts can interactively integrate information using an expressive first-order logic based language (WebPDDL) to define structure, semantics and mappings of data resources along with an inference engine.

All of the above efforts stop at the use of RDF and ontologies for raw data representation and capture of ground truth in the domain. The standard query mechanism supported over this RDF store is based on SPARQL query language [2]. The proposed GaBiD stack builds on the RDF representation as the first layer and creates appropriate query mechanism and graph and matrix based analytics over it to enable insight discovery and data exploration.

Multiple Big Data Stacks have been proposed with an intent of analyzing Big Data. The most popular one among them is the stack proposed by the Apache Hadoop project[8]. The Map Reduce framework and noSQL stores prescribed there are based on data representation in the form of key-value pairs. This representation is very good for data parallel applications but do not work well when the data elements are related to each other. They do not even capture explicit relationships between data. The proposed GaBiD stack leverages the implicit and explicit relationships between data elements and uses an intermediate graph representation for analyzing data.

4   Proposed Demonstration

The above described stack, semantic data modeling and graph-based visualization will be demonstrated in the conference. The business use case will be about analyzing Customer Care data to understand new patterns in data about problems reported by customers and the solutions employed by agents.

For the sake of simplicity, we consider data fusion from different database tables as inputs. The RDF model of the fused data is fed into a visualization toolset called Gephi [9]. Using a SPARQL query interface, user can select interesting sub-sets of the dataset to visualize. The user can then execute the required clustering algorithms from the panel and visualize with different Gephi supported layouts. Once an interesting graphical view

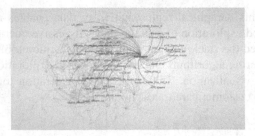

Fig. 2. Graphical Visualization   of Customer transactions

is generated , it can further be exported on to a web portal for offline exploration, supported by simple keyword-based queries. This makes visual exploration of large number of nodes feasible and appealing. Fig. 2 give some screenshots of the visualization.

5   Conclusions

In this demonstration paper, we described a new way of analyzing data coming from different sources. We proposed an alternate Big Data stack that utilized graph-based intermediate representation to derive new insights and provide intuitive visualization of data in Business domains. The proposed stack is applicable to multiple domains and will be demonstrated with customer care transaction data. In future, we plan to extend the stack and add new graph analysis algorithms and transformations to enable mining of implicit relationships between data as well.

References

1. Resource Description Framework, RDF, http://www.w3.org/RDF
2. SPARQL Query Language for RDF, http://www.w3.org/TR/rdf-sparql-query/
3. An ontology-based platform for semantic interoperability, Misikoff, Taglino. Springer (2004)
4. Cui, Z., Jones, D., et al.: Issues in Ontology-based Information Integration. In: IJCAI (2001)
5. Budgen, D., Rigby, M., et al.: A Data Integration Broker for Healthcare Systems. In: IEEE Computer 2007 (2007)
6. D2R, M.A.P.: – A Database to RDF Mapping Language, Christian Bizer. In: WWW 2003 (2003)
7. Dou, D., Pendu, P.L., et al.: Integrating Databases into the Semantic Web through an Ontology-based. In: ICDEW 2006 (2006)
8. Apache Hadoop Project, http://hadoop.apache.org
9. Gephi, The Open Graph Viz Platform, http://gephi.org

Processing Spatio-temporal Data
On Map-Reduce

Himanshu Gupta and Sriram Lakshminarasimhan

IBM India Research Laboratory
{higupta8,sriram.ls}@in.ibm.com

Abstract. The amount of spatio-temporal data generated in numerous scientific and industrial settings have exploded in recent years. Without a distributed platform, supporting efficient analytics operations over such voluminous datasets become prohibitively expensive. As a result there has been an increasing interest in using map-reduce to parallelize the processing of large-scale spatio-temporal data. While Hadoop, which has become the de-facto implementation of map-reduce, has shown to be effective in handling large volumes of unstructured data, several key issues needs to be addressed to exploit its power for processing spatio-temporal data.

In this tutorial, we explore design techniques for spatio-temporal analytics and data management on Hadoop, based on recent work in this area. We outline strategies for devising map-reduce algorithms for performing fundamental spatial analytics involving computational geometry operations as well as two-way and multi-way spatial join operations. We discuss storage optimization techniques such as chunking and colocation to enable efficient organization of multi-dimensional data on HDFS along with indexing techniques for fast spatial data access.

Keywords: spatio-temporal, map-reduce, computational geometry, spatial joins, spatial indexing.

1 Introduction

Spatio-temporal data arises in a number of scientific and industrial settings such as agriculture, climatology, seismology and telecommunications. Because of improved data acquisition capabilities, the amount of data that needs to be managed and analyzed can run into tens or hundreds of terabytes over time. The spatio-temporal analytics necessary to derive insight from these voluminous datasets are, in many cases, computationally intensive. Due to these reasons, there has been a surging interest in recent years in using map-reduce as a framework of choice for building distributed applications to process spatio-temporal data. The objective of this tutorial is to provide a detailed overview on how to efficiently distribute, query and perform analytics over spatio-temporal data on a cluster, using map-reduce.

S. Srinivasa and S. Mehta (Eds.): BDA 2014, LNCS 8883, pp. 57–59, 2014.

2   Tutorial Outline

The tutorial will start with then basics of Hadoop/HDFS/Map-Reduce [5] descibing the map-reduce platform and algorithmic design of parallel programs on map-reduce. We then detail some real-world use cases involving spatio-temporal data and analytics. We will then present how we can efficiently implement some of the commonly used computational geometric algorithms on map-reduce. Using these examples, we will discuss how we can design efficient algorithms for processing spatio-temporal data in particular and efficient map-reduce algorithms in general.

The tutorial will then discuss how we carry out joins on spatial and interval data efficiently. As intervals and spatial objects like polygons have a spatial extension, the algorithms for joining real-valued data can not be directly applied. Due to these reasons, joins on interval and spatial data is a non-trivial task. The tutorial will discuss both 2-way and multi-way joins on a series of predicates like *overlap, contains, before, range* etc. Through the help of join use-case we will illustrate a key map-reduce algorithm design principle that it is often more efficient to implement an analytics as two or more map-reduce jobs rather than as a single map-reduce job.

Finally, the tutorial will discuss various aspects related to storage and indexing of multi-dimensional data on map-reduce. We will discuss how these spatio-temporal datasets can be logically partitioned and physically colocated to improve the performance of map-reduce algorithms. We will also outline some techniques for indexing these spatio-temporal datasets on a distributed setting, in support of fast retrieval.

The detailed outline of our tutorial is given below.

1. Introduction
2. Motivating use-cases
3. Computational geometry operations [1]
    (a) Union
    (b) Skyline
    (c) Convex Hull
    (d) Closest-Pair/Farthest Pair
    (e) Voronoi Diagram
4. Two-way join processing
    (a) Spatial Predicates - Overlap, Range, Nearest-Neighbor
    (b) Interval Predicates - Overlap, Contains, Before
5. Multi-way join processing
    (a) Processing Multi-way Spatial Joins [4]
    (b) Processing Multi-way Interval Joins [2]
6. Storage
    (a) Chunking
    (b) Colocation [6]
7. Spatial access methods
    (a) Local and global indexing
    (b) R-Tree indexing
    (c) Geo-hashes
8. Conclusions

3   Duration

The preferred mode will be half-day tutorial.

4   Prerequisite Knowledge of Audience

Databases and a basic knowledge of map-reduce framework.

5   Tutorial History

This tutorial has not been presented else-where. To the best of our knowledge, no tutorial focused on processing spatio-temporal data on map-reduce has been presented at any venue. Afrati et al. [3] have a tutorial titled "Designing good algorithms for map-reduce and beyond" at SOCC-2012. This tutorial is focused on map-reduce algorithm design in general and not on spatio-temporal data processing on map-reduce.

6   Speaker Bios

Himanshu Gupta: is currently working as a technical staff member at Information and Analytics group at IBM Research, India. His research interests include information integration, hadoop and map-reduce processing, databases etc. He is currently working on various IBM initiatives which focus on processing spatio-temporal data on map-reduce. He holds a BTech and MS in Computer Science from IIT Kanpur and IIT Delhi respectively.

Sriram Lakshminarasimhan: is a Researcher in the Information and Analytics group at IBM India Research Laboratory. He received his PhD in Computer Science from North Carolina State University in Raleigh, USA in 2013. His research interest include high-performance computing, spatio-temporal data management and analytics, indexing and query processing.

References

1. Eldawy, A., et al.: CGHadoop: Computational Geometry in MapReduce. In: SIGSPATIAL (2013)
2. Chawda., B., et al.: Processing Interval Joins On Map-Reduce. In: EDBT (2014)
3. Afrati, F.N., et al.: Designing good algorithms for MapReduce and beyond. In: SOCC (2012)
4. Gupta, H., et al.: Processing Multi-way Spatial Joins On Map-Reduce. In: EDBT (2013)
5. Dean, J., et al.: MapReduce: Simplified data processing on large clusters. Comm. of ACM 51(1) (2008)
6. Eltabakh, M., et al.: CoHadoop: Flexible Data Placement and its exploitation in Hadoop. In: VLDB (2011)

Combining User Interested Topic and Document Topic for Personalized Information Retrieval

K. Veningston and R . Shanmugalakshmi

Department of Computer Science and Engineering,
Government College of Technology, Coimbatore, India
{veningstonk,cseit.gct}@gmail.com

Abstract. Personalization aims to improve user's searching experience by tailoring search results according to individual user's interests. Typically, search engines employ two-level ranking strategy. Firstly, initial list of documents is prepared using a low-quality ranking function that is less computationally expensive. Secondly, initial list is re-ranked by machine learning algorithms which involve expensive computation. The proposed approach explores the second level of ranking strategy which exploits user information. In this approach, queries and search-result clicks are used to model the user interest profiles probabilistically. The user's history provides the prior probability that a user searches for a topic which is independent of user query. The document topical features are combined with user specific information to determine whether a document satisfies user's information need or not. The probability of relevance of each retrieved document for a query is computed by integrating user topic model and document topic model. Thus, documents are re-ranked according to the personalized score computed for each document. The proposed approach has been implemented and evaluated using real dataset similar to AOL search log dataset for personalization. Empirical results along with the theoretical foundations of the model confirm that the proposed approach shows promising results.

Keywords: Information Retrieval, Personalization, Re-ranking, Probabilistic model, Topic model.

1   Introduction

Web Information Retrieval (IR) [4][27] process faces the problems of information mismatching and overcapacity. As the amount of information on the Web increases rapidly, it creates many new challenges for Web search. When the same query is submitted by different users, a typical search engine returns the same result, regardless of who submitted the query. However with the recent advent of click data [38], web search engine now personalize search results quite often. At the same time, user's current interest for the same query may be different at different times, different places. The current web search approach may not be suitable for users with different information needs. For example, upon the query "java", some users may be interested in

S. Srinivasa and S. Mehta (Eds.): BDA 2014, LNCS 8883, pp. 60–79, 2014.
© Springer International Publishing Switzerland 2014

documents dealing with "java" as "a programming language" while some other users may want documents relating to "an island of Indonesia". This kind of queries is referred to as an ambiguous query which mean to more than one category of results. For this kind of an ambiguous query, different users may have different search goals when they submit it to a search engine. However, it should not be treated as an ambiguous query. If it is possible for the search engine to derive user interests upon the query, then the user's intention becomes obvious. Personalized IR has become a promising area for disambiguating the web search query and therefore improving retrieval effectiveness by modeling the user profile by using his/her interests and preferences. While many search engines take advantage of information about people in general, or about specific groups of people, personalized search [17] depends on a user profile that is unique to the individual. Often short queries are ambiguous which provides very little information to a search engine on which the most relevant Web pages among millions need to be selected. A user profile can be used to supplement information about the search that is currently being represented by the query itself. This information can be used to narrow down the number of topics/contexts to be considered while retrieving the results. This increases the likelihood of including the most interesting results from the user's perspective. Typically the commercial search engine such as Google uses cookies and location information in order to personalize advertisements and most likely search results as well. The main contribution of the proposed approach is to exploit the topics in addition to that of treating whole document and search history towards personalizing search.

In this work, user's search history, which is kept in a log format recording which queries the user has made in the past and which results he/she has chosen to view is utilized. This could be an important form of search context for the following reasons. First, a user's background and interests can be learned from the user's search history (e.g., by looking at the topics covered by the past queries). For example, if there have been a lot of queries like "car racing" and "Porsche club", the user is probably interested in sports cars and "jaguar" is likely to mean the car. Second, from the users past (implicit) indication of document relevance, his/her reaction to the current retrieved documents implicitly provides an indication of his interests. For example, if the user searched with the same query "jaguar" before and clicked on Jaguar US's homepage link, with high confidence it can be predicted that the user would do it again this time, and it makes good sense to list that webpage in the top. Even when there is no exact occurrence of the current query in history, still some similar queries would be helpful.

2   Current Practice and Research

When search query is issued, most of the search engines return the same results irrespective of the users' interest because it lacks the existence of a semantic structure and hence it requires understanding of the information provided by the user. The following reasons accumulate complexity of the search process. The process of identifying intention of the user becomes difficult due to information available about user are very limited. At the same time, users do not wish to express their interest explicitly.

They want information instantaneously on supplying search query. Most of the users supply inaccurate input keyword query which is imprecise. They often under specify their true information needs. Thus query becomes ambiguous which needs to be understood by the retrieval system. Hence, personalization [29] strategy needs to be adopted in order to solve these problems faced by the retrieval system.

2.1   Related Work

- **Short-term and Long-term personalization:** Short term personalization [20] describes a personalized search based on the current user session. This approach is shown to improve retrieval quality. Long term personalization [24] describes a personalized search based on the entire history of user search in order to learn about the long term user characteristics.
- **Session based personalization:** Most of the personalized retrieval strategies do not distinguish between short term and long term user interests and make use of the whole search history to improve the search accuracy. Thus session based personalization [23] learns user interests by aggregating concept-based short terms identified within related search sessions.
- **Query ambiguity prediction:** Given an ambiguous query, it is either preferable to adapt the search result to a specific aspect that may be of the user's interest or to predict multiple aspects in order to maximize the probability that some query aspect is relevant to the user.
- **Implicit user modeling:** Typical retrieval systems lack user modeling and are not adaptive to individual users. Thus it is essential to infer a user's interest from the user's search context and use the inferred implicit user model for personalized search. In [37], the previous query has been exploited to enrich the current query and provide more search context to help disambiguation if two consecutive queries are related. This approach also infers user's interest based on the summaries of the viewed documents. The computed new user model is then be used to rank the documents with a standard information retrieval model.
- **Collaborative personalization:** In order to increase the user satisfaction towards online information search, search engine developers try to predict user preferences based on other user behavior. Thus, recommendations provided by the search engines may support users at some extent. Collaborative personalization attempts to better understand whether groups of people can be used to benefit from personalized search. The approach proposed in [15], combines individual's data with that of other related people to enhance the performance of personalized search. The use of group information for personalization is termed as groupization.
- **Search interaction personalization:** The approach presented in [8] incorporates user behavior data in order to improve the ordering of top results in real web search setting. It examines the alternatives for incorporating feedback into the ranking process and explores the contributions of search user feedback. The approach presented in [13] uses click-through information for improving web search ranking and it captures only one aspect of the user interactions with web search engines.

- **Ontology based personalization:** The approach presented in [2] attempts to personalize search results that involve building models of user context as ontological profiles by assigning implicitly derived interest scores to existing concepts in domain ontology and maintain the interest scores based on the user's ongoing behavior. This approach demonstrate that the semantic knowledge embedded in an ontology combined with long-term user profiles can be used to effectively tailor search results based on users' interests and preferences. However, changes in user profiles over time needs to be captured in order to ensure the incremental updates to the interest scores accurately reflect changes in user interests.

2.2   Problem Description

30 users including undergraduate and postgraduate students of Government College of Technology, Coimbatore, India performed the task of retrieving documents that satisfy their needs. They were given list of keyword queries to be searched using the typical search engine. Some of the keywords and its intention behind those keyword queries with respect to different users have been given in Table 1.

2.3   Proposed Approach Overview

As the key issue with the abundance of online information is to find relevant web documents, personalization of content is the key to address this issue. This paper presents user profile model which incorporates user intent by analyzing user's previous searches, issued queries and clicked information. It includes the modules as follows. (i) User profile information gathering, (ii) User topic modeling (iii) Matching of user topic model and document topic model to compute personalized relevance score for re-ranking the documents.

Table 1. Diverse interest of search users

Original query	Intention on relevant documents		
	User 1	User 2	User 3
World cup	Web pages mainly dealing with the foodball championship	Web pages mainly dealing with the ICC cricket world cup	Web pages mainly dealing with the T20 cricket world cup
India crisis	Web pages dealing with the economic crisis in India	Web pages dealing with the security crisis in India	Web pages dealing with the job crisis in India
Apple	Web pages on Apple store	Web pages on varieties of apple fruit	Web pages on Apple OS updates and downloads
The ring	Web pages about Ornament	Web pages about the horror movie	Web pages about circus ring show
Okapi	Pages related to animal giraffe	Pages related to okapi African luxury hand bags	Pages related to Information retrieval model BM25

3   Proposed User Topic Modeling for Personalized Search

Statistical language modeling for IR has emerged within the past several years as a new probabilistic framework for describing information retrieval processes. Language Modeling refers to the task of estimating a probability distribution that captures statistical regularities of natural language use. Applied to IR, language modeling refers to the problem of estimating the likelihood that a query and a document could have been generated by the same language model, given the language model of the document and with or without a language model of the query.

3.1   Probabilistic Approach for Personalization

In this work a probabilistic model [7][21][1] is used for predicting the relevance of a document to a specific user with respect to a query. The user representation corresponds to user-specific parameters for part of the model. The formalization assumes that there are only document-specific latent variables (i.e., document features), user specific latent variables (i.e., information need for the query), and combines them to determine whether a document's features satisfy the user's information need. The browsing history of users is obtained from the search logs and user profiles are generated from the browsing behavior.

3.2   User Profile Modeling

Personalization [25] aims to provide users with what they need without requiring them to ask for it explicitly. This means that a personalized IR system must somehow infer what the user requires based on either previous or current interactions with the user. θ_u is defined as a set of terms that the user has come across during the previous and current search and its probability of occurrence in user search session. The User Profile (*UP*) is built with the terms present in users search history. Typically user's search history comprises the queries and documents clicked. The system obtains information on the user this way and infers what are user's needs based on this information. In order to apply the personalization approach, the probability distribution $P(T_u|\theta_u,q)$ as the probability that when issuing a query q, a user u is seeking information on topic T_u. θ_u denotes the user-specific parameters i.e. user profile (*UP*) which possesses *terms* present in user search history. To obtain this conditional distribution, learn a user-independent language model $P(q|T)$ and a user specific prior probability of the topic, $P(T|\theta_u)$, and then apply Bayes' theorem as described in [7].

$$\theta_u = UP_{w_i}$$

$$UP_{w_i \in History(D)} = P(w_i) = \frac{tf_{w_i,D}}{\sum_{w_i \in D} tf_{w_i,D}} \tag{1}$$

Table 2. Sample user profile representation

w_i	$P(w_i)$
Computer	0.012
Datastructure	0.02
Programming	0.0011
Instruction	0.001
Algorithm	0.032
Analysis	0.004

Topic modeling [6] can be a good choice for IR based problems as low dimensional topical representation can well represent the user search preferences optimistically. Topical modeling using Latent Dirichlet Allocation (LDA) in [35] and Probabilistic Latent Semantic Analysis (pLSA) in [32][3] has been successfully applied. The sample user profile θu shown in Table 2.

3.3   Training for User Interested Topic Identification

The user topic model is trained on the Open Directory Project (ODP) corpus [5] [http://www.dmoz.org/]. In this work, topical categories from the top most level of the ODP are used. ODP screen shot shown in Fig. 1 includes 15 broad categories such as arts, games, home, health, etc. and its sub-categories.

Fig. 1. ODP main page [http://www.dmoz.org/]

From the user's search history, it is assumed that a user's click on a document is equated with the observation rel(d,q) = 1 otherwise, rel(d, q) = 0 when there is lack of a click. It is assumed that the user's intended topic T_u is equal to the topic of the document that they click on. Un-clicked pages were assumed as irrelevant to the user in [7]. But, it is not fair that treating un-clicked pages as irrelevant because it could be either relevant or irrelevant to the user. The problem of finding user's negative preferences from un-clicked documents that are considered irrelevant to the user has been addressed by exploiting Spy Naïve Bayes (SNB) classifier [36]. The conventional Naïve Bayes requires both positive and negative examples as training data, while SNB require only positive examples.

Algorithm 1. *Relevance_group_user* (document d, query q)

$V(q)$ = set of users who have previously searched for q
for each user v
if v has clicked on document d for q
  rel(d,q) = 1
else
  rel(d,q) = 0
number of users, who find d as relevant for q, $N = \sum rel(d,q)$

The probability of relevance obtained from the *relevance function* is biased towards the population of users those who usually search on the query using the search engine. Suppose $V(q)$ = $\{v\}$ be the set of users who have previously searched for query q and whose relevance feedback is used to train the *ranking function*. The probabilistic model explicitly takes these users' intended topics into consideration when interpreting the probability of relevance computed by the ranking function. rel(d,q) is the expected relevance with respect to the distribution of users who typically search for query q across all possible query intents. In order to avoid biasing, the modified rel(d,q) is estimated for each user as per the Algorithm 2.

Algorithm 2. *Relevance_individual_user* (query q, document d, user u)

D = set of documents retrieved for query q
for each document d in D
if document d is clicked by u for query q
  rel(d,q) = 1
else
  rel(d,q) = 0
frequency of q in d

In order to learn topics of interest from the users search profile and document topics, the following Algorithms 3 & 4 have been implemented.

Algorithm 3. Training_user_topic(q,d, θ_u)

Input: query q in search log, doc d clicked for q by user u, user_profile θ_u)
Output: Topics that are of interest for u
for each query q in user's search history
    for each document d clicked by the user
for each topic T in the ODP category
      compute T_u by $P(T \mid \theta_u, q) = \delta(P(T \mid q)) + (1 - \delta)(P(T \mid \theta_u)P(q \mid T)$

return (T_u)

Algorithm 4. Finding_document_topic(q,D)

Input: initial query q, retrieved documents D
Output: Topic of the documents retrieved
    for each document $d_i \epsilon D$ retrieved for the query q
for each topic T in the ODP category
      compute T_d by $P(T \mid d) = P(t_i \in d \mid T)P(T)$

return (T_d)

In order to compute T_d, choose a topic T according to a multinomial distribution conditioned on document d for each term t_i of document d in the training set. Then generate the word by drawing from a multinomial conditioned on topic T. In this way, documents can have multiple topics. Substituting the values of $rel(d,q)$, distribution of query topics, topic of each document, probability of relevance of a document with respect to a query for a specific user is estimated and thus personalized re-ranking is computed.

3.4   Exploiting User interest profile Model

An essential component of personalized search is learning user's interests. The search history for each user consists of the queries issued, the list of documents in the visible search results, and the list of documents clicked on by the user in response to each query. Since personalization in IR aims at enhancing user's knowledge by incorporating the user preferences and judgment into the retrieval models, the usage of implicit user interests and preferences has been identified so as to enhance current retrieval algorithms and anticipate limitations as World Wide Web (WWW) content keeps increasing, and user expectations keep growing and diversifying. Without requiring further efforts from users, personalization aims to compensate the limitations of user need representation formalisms such as the keyword-based or document-based representations. Thus, User profile is modeled using the initial results set for a given query, Titles of each documents initially presented, extracted terms from full text, extracted terms from snippets, whether URL is clicked previously or not and Dwell time i.e. time spent in each web pages that were clicked alrcady. The goal of user modeling for personalization system is to gain the capability to adapt specific search context of their preferences to better suit their needs.

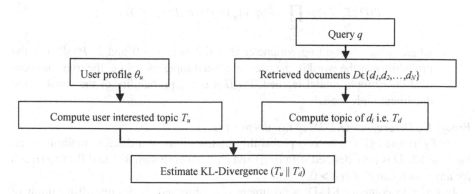

Fig. 2. User interested topic (T_u) vs. Document topic (T_d)

A model of information retrieval predicts and identifies what a user will find relevant given the user query. IR Models like Boolean model provides exact matching of documents and query. Vector space models consider the index representations as well as query as vectors embedded in a high dimensional Euclidean space, where each term is assigned a separate dimension. A Language Model (LM) refers to the task of estimating a probability distribution over all possible words in the document i.e. estimating the likelihood that a query and a document could have been generated by the same language model, given the language model of document and query. In this work, Probabilistic model proposed in [7] has been adopted to describe document's content by the topics. Also one more criterion is integrated into the basic retrieval model which is user profile represented by user interested topics. The discrete-valued variables T_d and T_u refer to the document's topic and the topic that the user is searching for, respectively. A document about topic T_d is assumed relevant to a user looking for topic T_u if the following points are met:

(1) Topic T_d satisfies users with information need T_u. Use Kullback-Leibler Divergence (KL-D) [28] between these two contextual models i.e. T_d and T_u to measure the similarity between two contexts. It is unlike exact matching instead matching based on the topicality of the information known already. KL-D metric quantifies the information gain between two probability distributions. It also measures the divergence of probability distribution of a topic in the document (T_d) to its distribution in the user interested topics (T_u). The lesser the divergence from T_u, the more informative the topic is for document T_d. The KL-D value of a topic t in document and user interest is given as shown in Eq. (2).

$$KL - D(T_d \parallel T_u) = \sum_{t \in D \cap U} P(T_d(t)) \log \frac{P(T_d(t))}{P(T_u(t))} \tag{2}$$

where $P(T_d(t))$ is the probability of topic t in document topic model T_d and $P(T_u(t))$ is the probability of topic t in user topic model T_u.

(2) Given that the document's topic matches that of the user's search intent i.e. user's topic, the document is relevant to the query based on Eq. (3).

$$P(Q \mid T_u, T_d) = \prod_{q_i \in Q} (\alpha P(q_i \mid T_u) + (1 - \alpha) P(q_i \mid T_d)) \tag{3}$$

where, α is a weighting parameter that lies between 0 and 1, $P(q_i|T_u)$ is the probability of the word in the user interested topic model i.e. the user independent query topic model learned, $P(q_i|T_d)$ is the probability of the word in the document topic model.

Proof: KL Divergence (KL-D) of two documents
Let $P(T_d(t))$ and $P(T_u(t))$ be two probability distributions of a discrete random variable. The KL-D is only defined if $P(T_d(t))$ and $P(T_u(t))$ both sum to 1 and if $P(T_u(t)) > 0$ for any t such that $P(T_d(t)) > 0$.

In order to compute KL-D, a document d is observed as discrete distribution of $|d|$ random variables, where $|d|$ is the number of words in the document. Let d_1 and d_2 be two documents for which we want to calculate their KL-divergence. It is run into two problems:

— Compute the KL-divergence twice due to asymmetry: KL-D($T_d \| T_u$) and KL-D($T_u \| T_d$).
— Due to the constraint for defining KL-D, calculations must only consider words occurring in both d_1 and d_2.

3.5   Personalized Re-ranking Process

Typically, there are two variants of user context information [37] to model user's search experience. Firstly, the short-term context which emphasizes that the most recent search is most directly close to the user's current information need. Successive searches in a session usually have the same context. However, detecting a session is a difficult task. Secondly, the long-term context which assumes that user will have their interests over a long time. It means that the past search may have some impact on current search. Re-ranking of the results reflects the most relevant results for the user. It is a process of re-ordering the retrieved results based on combination of short-term and long-term user preferences. Re-ranking computation performs the following two processes. They are,

— Calculating personalized score for document
— Generating personalized result set

3.5.1   Personalized Score

The personalized relevance score for each document d for a query q is computed for a user u who issued query q as follows:

— Compute the topic T_u of user interest
— Retrieve all the documents $d_1, d_2, \ldots\ldots, d_N$ for query q from a traditional search engine
— Compute the topic T_d of each document

The conditional distribution $P(T|d)$ specifies the topic of each document. This distribution could be estimated using techniques described in [36][32][3] in order to predict the ODP category for each web document. Typically, for the given query q and the user profile θ_u, in order to find the relevant documents, the documents are re-ranked by $P(D|Q,\theta_u)$ using Bayes' theorem as shown in Eq. (4).

$$P(D \mid Q, \theta_u) = \frac{P(D \mid \theta_u) P(Q \mid D, \theta_u)}{P(Q \mid \theta_u)} \tag{4}$$

The personalized score is computed for each user incorporating his/her interest and preferences. Let D be the set of documents returned by the search engine. The rank of each document D returned for a query Q for user u is computed by integrating topic model and user model as shown in Eq. (5).

$$P(Q \mid D, \theta_u) = P(Q \mid T_d, T_u) + \prod_{q_i \in Q} (\beta P(q_i \mid \theta_u) + (1-\beta) P(q_i \mid D)) \tag{5}$$

where, β is a weighting parameter that lies between 0 and 1, $P(q_i|\theta_u)$ is the probability of the word from the user interest profile model, $P(q_i|D)$ is the probability of the word from the documents retrieved i.e. document model.

3.5.2   Personalized Result

Search engines always return millions of search results and thus it is essential to re-order results to facilitate users to find documents what they want. Re-ordering web search results is assumed as an application of user interest modeling. The initial documents retrieved for the query by the search engine can be re-ranked according to the personalized score computed. The documents are then scored based on the probability $P(Q|D,\theta_u)$ and arranged based on descending order of the personalized score.

4   Experimental Setup

4.1   Dataset Description

The dataset used in this work is similar to AOL search log [12] which possesses implicit feedback in the form of click-through data collected by a search engine and it is released for research purpose. The AOL Search Data is a collection of real query log data that is based on real users. The data set consists of 20M web queries collected from 650k users over the period of three months.

Table 3. Statistics about AOL search log dataset

Number of lines of data	36,389,567
Number of instances of new queries (with or without click-through data)	21,011,340
Number of requests for "next page" of results	7,887,022
Number of user click-through data	19,442,629
Number of queries without user click-through data	16,946,938
Number of unique queries	10,154,742
Number of users log	657,426

4.2   Baseline Approaches

The following approaches have been considered for comparing the proposed personalization model to assess the performance improvements.

Best Matching 25 (BM25). BM25 (Best Matching) [31] is a bag-of-words retrieval function that ranks a set of documents based on the query terms appearing in each document, regardless of the inter-relationship between the query terms within a document. Given a query Q, containing keywords $q_1, q_2,...,q_n$, the BM25 probabilistic ranking function of a document D is computed as shown in Eq. (6).

$$score(D,Q) = \sum_{i=1}^{n} IDF(q_i) \cdot \frac{tf_i \cdot (k_1 + 1)}{tf_i + k_1 \cdot \left(1 - b + b \cdot \frac{|D|}{avgdl}\right)} \tag{6}$$

where, tf_i is q_i's term frequency in document D, $|D|$ is the length of the document D in terms of number of words, $avgdl$ is the average document length in the dataset. k_1 and b are tuning parameters. k_1 has little effect on retrieval performance, b is a document length normalization parameter to be ranging within 0-1.

$$IDF(q_i) = \log \frac{N - df_i + 0.5}{df_i + 0.5} \tag{7}$$

where, N is the total number of documents in the dataset and df_i is the number of documents containing the query q_i.

Rocchio Algorithm. This approach uses relevance feedback [14] to improve retrieval performance. Rocchio algorithm incrementally modifies the query by adding terms from the explicit relevance feedback. The typical Rocchio approach has been modified for comparison for work in personalized search. In order to construct the user profile, past queries Q_n entered by the user and its associated clicks as relevance feedback RF_n has been used. Thus, terms from $RF_{1...n}$ is extracted to compute their frequencies, which in turn represented as $\{(w_1, tf_{w_1}), (w_2, tf_{w_2}), ..., (w_i, tf_{w_i}), ..., (w_n, tf_{w_n})\}$ where w_i is a word in RF_n and tf_{w_i} is the frequency of occurrence of w_i. Then, the set of documents D returned by the search engine for a query Q is re-ranked by the personalized score computed as shown in Eq. (8).

$$Score(D, Q) = \sum_{w \in Q} \left(\frac{tf_{w,Q}}{|Q|} + \frac{tf_{w,UP}}{|UP|} \right) \cdot \frac{tf_{w,D}}{|D|} \tag{8}$$

where $tf_{w_i,Q}$ is the term frequency of word w_i in query Q, $tf_{w_i,UP}$ is the term frequency of w_i in user profile UP, $tf_{w_i,D}$ is the term frequency of w_i in document D, and $|Q|$, $|UP|, |D|$ are the length of the query i.e. number of words in Q, length of the user profile i.e. number of words in UP, length of the document i.e. number of words in D respectively. Thus, the documents retrieved for the initial query are then re-ordered based on the score computed.

Document Language Model (LM) Approach. In exploiting LM [37][19], the user profile is learned by collecting words and their probabilities from the implicit relevance feedback for all the training queries. In order to re-rank the retrieved documents, retrieve top few results from the traditional search engine and then re-order them based on the score computed as given in Eq. (9).

$$P(Q \mid D, UP) = \prod_{q_i \in Q} (\alpha P(q_i \mid UP) + (1 - \alpha) P(q_i \mid D)) \tag{9}$$

where $P(q_i|UP)$ is the probability of the word q_i in user profile, $P(q_i|D)$ is the probability of the word in the document and α is a weight tuning parameter which takes values between 0 and 1.

Query Language Model Approach. In this approach, the probability of the word in user profile is smoothed with a general LM estimated from a large number of queries from the query log as given in Eq. (10).

$$P(q_i \mid UP) = \beta P(q_i \mid UP) + (1 - \beta) P(q_i \mid QueryLog) \tag{10}$$

where $P(q_i | QueryLog)$ is the probability of the word q_i in the search query log and β is a weight tuning parameter which takes values between 0 and 1.

4.3   Evaluation Metrics

The re-ranking algorithms proposed in this work have been evaluated using a variety of accepted IR metrics [4][27].

— **Precision:** This measures the accuracy of the retrieved results. Precision defines the fraction of retrieved documents that are labeled as relevant i.e. documents ranked in the top n results that are found to be relevant. If the documents within the top k are irrelevant, then this measures the user satisfaction with the top k results.

$$P@k = \frac{\#of\_relevant\_doc\_retrieved\_among\_k}{k} \tag{11}$$

— **Recall:** This measures the coverage of the relevant documents in the retrieved results. Recall defines the fraction of relevant documents that are retrieved.

$$R@k = \frac{\#of\_relevant\_doc\_retrieved\_among\_k}{total\,\#of\_relevant\_documents} \tag{12}$$

— **Interpolated precision:** The interpolated precision ($P_{Interpolated}$) at a certain recall level r is defined as the highest precision found for any recall level $r' \geq r$.

$$P_{Interpolated}(r) = \max_{r' \geq r} P(r') \tag{13}$$

— **Mean Reciprocal Rank (MRR):** The mean reciprocal rank is the average of the reciprocal ranks of results for a sample of queries Q. Consider the rank position k of first relevant document, Reciprocal Rank score $=1/k$. MRR is the mean of reciprocal rank across multiple queries given by Eq. (14).

$$MRR = \frac{1}{|Q|} \sum_{i=1}^{|Q|} \frac{1}{rank_i} \tag{14}$$

— **Normalized Discounted Cumulative Gain (NDCG):** Precision and Mean Average Precision can only handle cases with binary judgment i.e. relevant or irrelevant. To measure the ranking quality accurately, Discounted Cumulative Gain (DCG) [18] has been used. DCG is a measure that gives more weight to higher ranked documents by discounting the gain values for lower ranked documents. This measure the usefulness of a retrieved document based on its position in the result list. The ranked results are examined from top ranked results to lower for a given query. The highly relevant documents appearing lower in search result list will be penalized by reducing relevance value r_i logarithmically proportional to the position of the result. The DCG accumulated at a particular rank position N is defined as given in Eq. (15).

$$DCG_K = \sum_{i=1}^{K} \frac{2^{r_i} - 1}{\log_2(i+1)} \tag{15}$$

where r_i is an integer relevance label (0= "Bad" and 5= "Perfect") of result returned at position i. "Bad" documents do not contribute to the sum, thus will reduce DCG for the query pushing down the relevant labeled documents, reducing their contributions. Since search result lists vary in length depending on the query issued to the search engine, the results of one query cannot be consistently compared with the results returned to another query using DCG measure. In order to normalize DCG, sort documents of a result list by the order of relevance to produce the maximum possible DCG till the position K i.e. termed as ideal DCG (IDCG). The normalized DCG is computed using Eq. (16).

$$NDCG_K = \frac{DCG_K}{IDCG_K} \tag{16}$$

5   Result Analysis

5.1   Experimental Design

The data for each user consists of queries and their corresponding clicked URLs. The training data is used for learning user profile and testing data is used for evaluating the approaches.

Training: The search results returned by the traditional search engine for the past queries are collected. The corresponding top documents/snippets from the search engine are extracted using a plug-in and then used for learning user profile.

Testing: The top N results returned by the traditional search engine are collected. These results are then passed to the personalized re-ranking process. This process then re-scores the results and returns the re-ranked results. The top N results from the re-ranked results are compared with the relevance judgment. Thus, the performance of the re-ranking approach is evaluated.

The AOL dataset contains only the URL of clicked documents. It does not contain the actual document content. Hence, the real dataset has been generated and used for the experimental evaluation. The experiments were carried out on a dataset consisting of 50 users. Each user was asked to submit number of queries to traditional search engine. For each query, they were further asked to examine the top 20 results in order to identify the set of relevant documents.

Table 4 gives the sample statistics of the real dataset of 10 users. The document collection consists of top 20 documents for each query from the search engine which is typically assumed to be assessed by the user while retrieving the relevance of the documents. The total size of the document collection is 14265 documents excluding '.ppt', '.pdf' and '.doc' formats. The commercial search engine is used in order to retrieve initial set of results matching the query. The proposed implementation is simulated using Lucene based IR system (http://lucene.apache.org/).

In order to construct the user profile, divide the search history of each user into two groups in such a way that each of which possesses equal number of search queries. For example, User 1 searched for 43 different queries, divide his/her search log into 2 sets with 21 queries each approximately. Then, user profile is learned for first set. The second set is used for testing purpose.

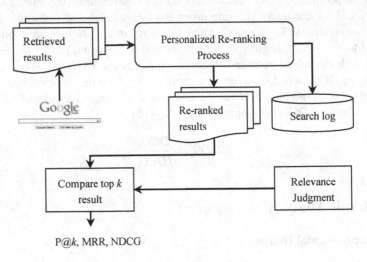

Fig. 3. Experimental setup for evaluation

In order to learn the user interest profile, the past search queries issued and its corresponding relevant documents are assumed. Performance is calculated during testing. In testing, user was asked to enter the query. Subsequently, re-ranked set of documents is generated using the proposed approach using user profile learned from first set of queries. Accordingly, precision at top k, MRR and NDCG are measured to show the performance improvement over baseline systems.

Table 4. Sample real dataset statistics (10 Users)

User	# of Queries	Total # of relevant documents	Average # of relevant documents
User 1	43	225	5.23
User 2	39	125	3.21
User 3	63	295	4.68
User 4	62	188	3.03
User 5	37	190	5.14
User 6	28	91	3.25
User 7	45	173	3.84
User 8	31	96	3.10
User 9	51	240	4.71
User 10	39	128	3.28

5.2   Parameter Tuning

The performance of the proposed method is sensitive to the choice of α and β parameters. These parameters needs to be tuned as there are two ranking combination schemes shown in Eq. (3) and Eq. (5) to be used. The parameter α determines the weight of the user interested topic model (T_u) and document topic model (T_d) while β parameter determines the weight of the user interest profile model (θ_u) and document model (D). Taking the top 10 search results as an instance, we give a range of values for α and β and compare the relative improvement in Precision, MRR, and NDCG. We compare the two ranking combination schemes and the results are shown in Fig. 4 and Fig. 5 respectively. With regard to the proposed scheme, as long as α and β value is big enough, the improvement in IR measures stay around the maximum value without much diminishing change. Although the optimum value of α and β is hard to formulate, the empirical results show that if we simply re-rank totally by user interest profile model (θ_u) and user interested topic model (T_u), the improvement in Precision, MRR, and NDCG is very close to the maximum value that can be achieved.

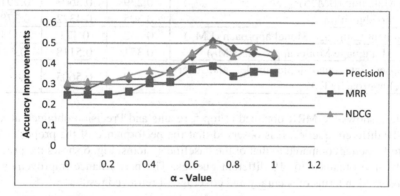

Fig. 4. α Parameter tuning on Eq. (3) for 5 queries at top 10 search results

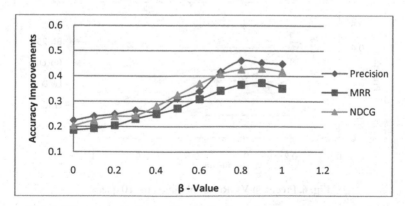

Fig. 5. β Parameter tuning on Eq. (5) for 5 queries at top 10 search results

From the Fig. 4 and Fig. 5, it is noted that the results of the combination of different models is getting better when the values of α and β is sufficiently large. Initially β-value was set 0.5 while tuning the α-value and then α-value was set 0.7 while tuning the β-value. Thus, it is implied that α and β values 0.7 and 0.85 respectively yields better accuracy on an average.

5.3   Evaluation on Real Dataset

Experiments have shown that the proposed personalization approach (T_U) achieve better results over baseline methods. The rich representation of user interest model served as a fine user search context model which capture user search goal accurately. Thus, bring the relevant documents in the first few results.

Table 5. MRR and Precision at top-k results for 10 queries

Method	MRR@5	P@5	P@10
Best Matching (BM25)	0.239	0.3607	0.2914
Rocchio algorithm	0.305	0.4322	0.3783
Document Language Model approach (LM_D)	0.332	0.473	0.4145
Query Language Model approach (LM_Q)	0.371	0.5118	0.447
Proposed integrated topic model and user model approach (T_U)	0.428	0.5605	0.4926

Table 5 shows the MRR obtained at top 5 results and Precision obtained at k ∈ 5, 10 for 10 different queries. It is observed that the performance of the proposed T_U is found to be better compared to that of the baseline systems. Fig. 6 shows the precision - recall curve obtained for 10 different queries. The performance improvement has been observed in terms of accuracy and coverage of retrieved results.

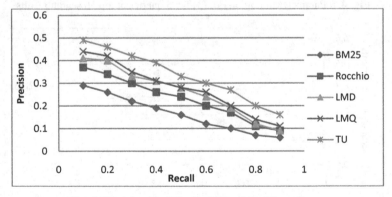

Fig. 6. Precision Vs. Recall obtained for 10 queries

Fig. 7 shows NDCG obtained at k ∈ 1, 2, 3,..., 10 for 10 different queries. It is observed that the performance of the proposed TU is found to be consistently better compared to that of the baseline systems. It is observed that the proposed method shows performance improvement over baseline methods on real dataset by bringing the highly relevant documents in first few results. Thus, increased NDCG values at first 10 results means that the proposed method ranks documents appropriately incorporating user search context in order to satisfy users with relevant documents.

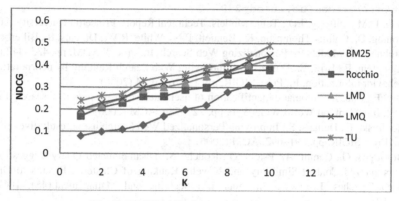

Fig. 7. NDCG at K obtained for 10 queries

6   Conclusion

Personalization has been performed at client side by re-ranking the results returned by the traditional search engine. The distribution of user topics and probability distribution of document topics have been estimated to calculate the personalized score in order to observe the relevance of documents. The documents are then re-ranked according to the score obtained. There is merit in locally generating a topic model, and then locally filtering and re-ranking search results, as this approach can work even when cookies are not accepted or deleted by the browser to maintain privacy. However, it is inferred that still there is a gap existing in the process of user profile information capture and representations.

Acknowledgements. The work presented in this paper is supported and funded by the Department of Science and Technology (DST), Ministry of Science and Technology, Government of India under INSPIRE scheme. Authors wish to extend their thanks to DST.

References

1. Berger, A., Lafferty, J.: Information retrieval as statistical translation. In Proc. SIGIR, pp. 222–229. ACM (1999)
2. Sieg, A., Mobasher, B., Burke, R.: Web search personalization with ontological user profiles. In: Proc. CIKM, pp. 525–534. ACM (2007)

3. Lin, C., Xue, G.-R., Zeng, H.-J., Yu, Y.: Using Probabilistic Latent Semantic Analysis for Personalized Web Search. In: Zhang, Y., Tanaka, K., Yu, J.X., Wang, S., Li, M. (eds.) APWeb 2005. LNCS, vol. 3399, pp. 707–717. Springer, Heidelberg (2005)
4. Manning, C.D., Raghavan, P., Schutze, H.: Introduction to Information Retrieval. Book published by Cambridge University Press (2008)
5. Carpineto, C., Romano, G.: ODP239 dataset (2009), http://credo.fub.it/odp239/
6. Blei, D.M., Lafferty, J.D.: Topic Models. Technical Report, Princeton University (2009)
7. Sontag, D., Collins-Thompson, K., Bennett, P.N., White, R.W., Dumais, S., Billerbeck, B.: Probabilistic Models for Personalizing Web Search. In: Proc. WSDM, pp. 433–442 (2012)
8. Agichtein, E., Brill, E., Dumais, S.: Improving Web Search Ranking by ncorporating user behavior information. In: Proc. SIGIR, pp. 19–26. ACM (2006)
9. Qiu, F., Cho, J.: Automatic identification of user interest for personalized search. In: Proc. 15th Intl. Conf. on World Wide Web, pp. 727–736. ACM (2006)
10. Radlinski, F., Dumais, S.: Improving Personalized Web Search using result diversification. In: Proc. SIGIR, pp. 691–692. ACM (2006)
11. Bordogna, G., Campi, A., Psaila, G., Ronchi, S.: Disambiguated Query Suggestions and Personalized Content-Similarity and Novelty Ranking of Clustered Results to Optimize Web Searches. Elsevier - Information Processing and Management (48), 1067–1077 (2012)
12. Pass, G., Chowdhury, A., Torgeson, C.: A Picture of Search. In: Proc. 1st Intl. Conf. on Scalable Information Systems (2006)
13. Xue, G.-R., Zeng, H.-J., Chen, Z., Yu, Y., Ma, W.-Y., Xi, W., Fan, W.: Optimizing web search using web click-through data. In: Proc. CIKM, pp. 118–126. ACM (2004)
14. Rocchio, J.J.: Relevance feedback in information retrieval. In: Proc. The smart retrieval system - Experiments in Automatic Document Processing, pp. 313–323 (1971)
15. Teevan, J., Morris, M.R., Bush, S.: Discovering and using groups to improve personalized search. In: Proc. WSDM, pp. 15–24. ACM (2009)
16. Teevan, J., Dumais, S.T., Liebling, D.J.: To Personalize or Not to Personalize: Modeling Queries with Variation in User Intent. In: Proc. SIGIR, pp. 163–170. ACM (2008)
17. Teevan, J., Dumais, S.T., Horvitz, E.: Beyond the Commons: Investigating the Value of Personalizing Web Search. In: Proc. Workshop New Technologies for Personalized Information Access (PIA), pp. 84–92 (2005)
18. Kalervo, J., Jaana, K.: Cumulated Gain-Based Evaluation of IR Techniques. ACM Transactions on Information Systems (2002)
19. Ponte, J.M., Croft, W.B.: A language modeling approach to information retrieval. In: Proc. SIGIR, pp. 275–281. ACM (1998)
20. Hu, J., Chan, P.K.: Personalized Web Search by Using Learned User profiles in re-ranking. In: Proc. SIGKDD, pp. 1–14. ACM (2008)
21. Gao, J., He, X., Nie, J.-Y.: Clickthrough-Based Translation Models for Web Search: from Word Models to Phrase Models. In: Proc. CIKM, pp. 1139–1148. ACM (2010)
22. Sugiyama, K., Hatano, K., Yoshikawa, M.: Adaptive Web Search Based on User Profile Constructed without any Effort from Users. In: Proc. 13th Intl. Conf. on World Wide Web, pp. 675–684. ACM (2004)
23. Daoud, M., Tamine-Lechani, L., Boughanem, M.: Learning user interests for a session-based personalized Search. In: Proc. 2nd Intl. Symposium on Information Interaction in Context, pp. 57–64. ACM (2008)
24. Matthijs, N., Radlinski, F.: Personalizing Web Search using Long Term Browsing History. In: Proc. WSDM, pp. 25–34. ACM (2011)

25. Agrawal, R., Gollapudi, S.: Diversifying Search Results. In: Proc. WSDM, pp. 5–14. ACM (2009)
26. Krestel, R., Fankhauser, P.: Reranking web search results for diversity. Springer Information Retrieval (15), 458–477 (2012)
27. Baeza-Yates, R., Ribeiro-Neto, B.: Modern Information Retrieval. Addison Wesley (1999)
28. White, R.W., Chu, W., Hassan, A., He, X., Song, Y., Wang, H.: Enhancing personalized search by mining and modeling task behavior. In: Proc. WWW, pp. 1411–1420. ACM (2013)
29. Anand, S.S., Mobasher, B.: Intelligent techniques for web personalization. In: Mobasher, B., Anand, S.S. (eds.) ITWP 2003. LNCS (LNAI), vol. 3169, pp. 1–36. Springer, Heidelberg (2005)
30. Stamou, S., Ntoulas, A.: Search Personalization through Query and Page Topical Analysis. Proc. User Model User-adapt Interact 19(1-2), 5–33 (2009)
31. Robertson, S.E., Walker, S., Hancock-Beaulieu, M., Gull, A., Lau, M.: Okapi at TREC. In: Proc. Text Retrieval Conference, pp. 21–30 (1992)
32. Hofmann, T.: Probabilistic Latent Semantic Indexing. In: Proc. SIGIR, pp. 50–57. ACM (1999)
33. V.: K, M. Simon. Collaborative Filtering for Sharing the Concept Based User Profiles. In: Proc. of 3rd IEEE International Conference on Electronics Computer Technology (ICECT), vol. 4, pp. 187–191 (2011)
34. Veningston, K., Shanmugalakshmi, R.: Enhancing personalized web search re-ranking algorithm by incorporating user profile. In: Proc. of 3rd IEEE International Conference on Computing Communication and Networking Technologies (ICCCNT), pp. 1–6 (2012)
35. Wei, X., Bruce, C.W.: LDA-based document models for ad-hoc retrieval. In: Proc. SIGIR, pp. 178–185. ACM (2006)
36. Ng, W., Deng, L., Lee, D.L.: Mining User Preference Using Spy Voting for Search Engine Personalization. ACM Trans. Internet Technology 7(4), article 19 (2007)
37. Shen, X., Tan, B., Zhai, C.: Implicit User Modeling for Personalized Search. In: Proc. CIKM, pp. 824–831. ACM (2005)
38. Dou, Z., Song, R., Wen, J.-R., Yuan, X.: Evaluating the Effectiveness of Personalized Web Search. IEEE Trans. Knowledge and Data Engineering 21(8), 1178–1190 (2009)

Efficient Implementation of Web Search Query Reformulation Using Ant Colony Optimization

K. Veningston and R. Shanmugalakshmi

Department of Computer Science and Engineering,
Government College of Technology, Coimbatore, India
{veningstonk,cseit.gct}@gmail.com

Abstract. Typically, web search users submit short and ambiguous queries to search engines. As a result, users spent much time in formulating query in order to retrieve relevant information in the top ranked results. In this paper, term association graph is employed in order to provide query suggestion by assessing the linkage structure of the text graph constructed over a collection of documents. In addition to that, a biologically inspired model based on Ant Colony Optimisation (ACO) has been explored and applied over term association graph as learning process that addresses the problem of deriving optimal query suggestions. The user interactions with the search engine is treated as an individual ant's navigation and the collective navigations of all ants over the time result in strengthening more significant paths in a term association graph which in turn used to provide query modification suggestions. We present an algorithm that attempts to select the best related keyword among all possible suggestions for an input search query and discuss its implementation based on a ternary search tree and graph data structure. We experimentally study the performance of the proposed method in comparing with different techniques.

Keywords: Internet of things, Information Retrieval, Query refinement, Graph and trie representation, web search, Personalization.

1 Introduction

Search engines have become much more interactive in recent years which activated a lot of work in automatically acquiring knowledge structures that can assist users in navigating through a large document collection. Query log analysis has emerged as one of the most promising research areas to automatically derive such knowledge structures. Typically, Information Retrieval (IR) [23][3] has become one of the dominant area of research in web mining due to the growth and evolution of web documents. When the query is submitted by a user, a typical search engine returns a large set of results. Users may be expecting relevant documents in the first few pages of search results for the query. Thus, user often modify previously issued search query with an expectation of retrieving better results. These modifications are called query reformulations or query refinements. Query reformulation aims to solve the vocabulary mismatch problem in IR by changing the original query to form a new query that

S. Srinivasa and S. Mehta (Eds.): BDA 2014, LNCS 8883, pp. 80–94, 2014.
© Springer International Publishing Switzerland 2014

would find better relevant documents. Query refinement comprises query recommendation, suggestion, modification or expansion [16]. Suggestion of alternative terms to refine user's query is an effective technique to help the user quickly narrow down his/her specific information need.

2   Related Work

In order to satisfy the information needs of WWW users and improve the user experience especially in search task, different approaches have been studied. The idea of assisting web search users in the retrieval process by providing interactive query modifications has been extensively studied in [10]. It is apparent that the queries entered in typical IR systems by users are very short normally formed with not more than two or three words [17] which may not avoid the inherent ambiguity of language i.e. synonymy, polysemy, etc. Thus, query suggestions can assist users in search process [18]. In order to recommend relevant queries to Web users, query suggestion technique has been employed by current search engines such as Google, Yahoo!, and Ask. The aim of query suggestion is similar to that of query expansion [20][15], query substitution [24], and query refinement [21]. All these approaches focus on understanding users search intentions and improving the queries submitted by the users. With the increase in usage of Web search engines, it is easy to collect and use user query logs. The system developed in [5] extracts probabilistic correlations between query terms and documents terms using query logs. These correlations are then used to find good expansion terms for new queries. An approach presented in [6] assumes the documents visited by users are relevant to the query. It maintains a list of all the documents visited for a particular query. Probability of document being visited when a particular query word is present in a query is calculated to find the relevance of the document. The larger query log improves the retrieval accuracy of the system. The work presented in [26] considers query log as bipartite graph that connects the query nodes to the URL nodes by click edges. Given a query node q and a URL node u, there will be an edge (q,u) if u is the clicked answer for query q. The weight of the edge (q,u) is computed by the click frequency. Random walk [19] probability is used to find n most similar queries to q and combine them to form an expanded query. Typically query log analysis is performed in order to optimize Web search results ranking [11][25]. Web search logs are widely used to effectively organize the search results. In addition to that, a ranking function is learned from the implicit feedback extracted from search engine click-through data to provide personalized search results for users. ACO has been employed in order to learn adaptive knowledge structures from query logs in [7].

3   ACO Approach for Personalized Query Reformulation

The proposed approach is inspired from the social behaviors of ant in nature called ACO [8][9]. Recently ACO has been applied to learn adaptive knowledge structures from query logs [1][7]. However, adaptive knowledge structures are essentially difficult to assess. In this approach, ACO is explored to build adaptive knowledge structures for query reformulation in order to provide interactive search. The proposed approach promote semantic search that improves search accuracy by understanding user intent and the contextual meaning of the terms appear in the document corpus. Semantic search systems consider various features including search contexts, location, intent, synonym, polysemic words, generalized and specialized queries, concept matching and natural language queries in order to provide relevant search results to the user. The main idea is that the self-organizing principles which allow the highly coordinated behavior of real ants are exploited to coordinate populations of artificial agents that collaborate to solve computational problems.

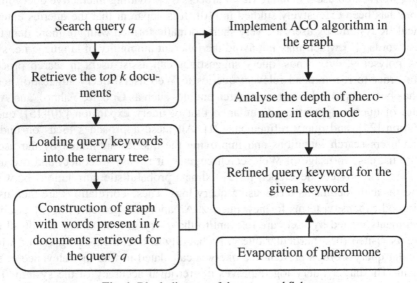

Fig. 1. Block diagram of the proposed Scheme

3.1   IR Enabled with Ant System

— The first ant finds the related keywords (k_i) via any path (p) in the graph, then returns to the query keyword (q), leaving behind a pheromone (w) trails i.e. weight or importance of the path.
— Ants arbitrarily follow various paths in the term graph. This in turn strengthens the path and makes it more attractive as an important route.
— Ants take the important route to find the related keyword which retrieve relevant documents and leaves the other paths.

3.2   ACO for Search Query Reformulation

Ternary Tree Construction. A Ternary Search Tree (TST) is a special trie data structure where the child nodes of a standard trie are ordered as a binary search tree. The Fig. 2 shows an example ternary search tree with the strings "CAT", "BUG", and

"UP" and its associated graph structure. At the end of each keyword in ternary tree, a term graph is connected. Term graph is a graph in which terms $(t_{1,...,n})$ present in top-k documents retrieved for the keyword query are interlinked based on its presence in the document. The TST has been used in [2] for correcting frequent spelling errors in user queries.

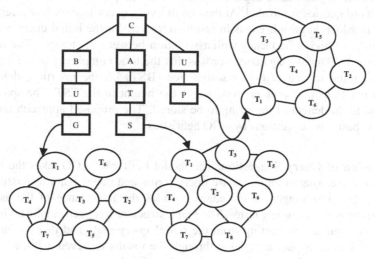

Fig. 2. Ternary tree nodes and its associated Term Graph

Ternary search trees are efficient to use in terms of space when the strings to be stored share a common prefix. Set of queries Q $(q_1, q_2,..., q_n)$ are fed into ternary tree and query node will act as a root node in the graph to be constructed in next phase.

Term Graph Construction. Documents retrieved for the q_i are preprocessed and terms are extracted in order to construct a graph. The term graph has been modeled based on the Algorithm 1.

Algorithm 1. Term_Graph_Construction (q_i, D_n)

Input: query term q_i, terms in retrieved documents d_t
Output: Graph structure G_d
      for each document $d_i \epsilon D$ retrieved for the query q
          extract terms present in d_i and create node for each term and create
          edges   between these nodes to be complete graph;
        for each term d_t
         if a term d_t present in more than one document then
         create an edge connecting d_t with terms present in documents

---

3.3 Query Reformulation Phase

Generation of Candidate Suggestions. The data set contains computer science related keywords and the keywords rating. Keyword rating is measured as the frequency of a word in resultant documents of the input query among group of users. Rating is considered as a pheromone depth or weight in the graph. TST is constructed and all the keyword queries are loaded. At the end of every query keyword, adjacency list is created in which keywords exists in results retrieved for the initial query are entered in the order of higher frequency with association between the terms. The reason for employing TST for storing query words is that the time complexity of the TST operations is similar to that of binary search tree (BST) i.e. the insertion, deletion and search operations take time proportional to the height of the TST. The space is proportional to the length of the string to be stored. The proposed approach attempts to select the best few suggestions by ACO heuristics.

Preparation of Query Suggestion. The model built with ACO takes the form of a graph structure where root nodes are query terms and edges point to possible query refinements. The weight on the edges encodes the importance of the association between the nodes i.e. query terms. The graph structure has been employed in order to recommend queries by starting from the initial query node and then to traverse the graph by simulating user navigation through the results retrieved and thus following ACO principle in order to identify and rank the related query suggestions.

3.4 Steps in Personalized Query Reformulation Using ACO

— Select a vertex $r \epsilon V[G]$ to be a root vertex that is query node
— Traverse the graph according to the navigation history of different user search
— Choose the list of vertices visited by most of the users from the query node
— Return the top few words visited as suggestions

Almost all ants are assumed to be blind and incapable of achieving complex tasks alone. It relies on the phenomena of swarm intelligence for survival and it is capable of establishing shortest-route paths from their colony to feeding sources and back. It uses stigmergic communication via pheromone trails and follows existing pheromone trails with high probability. Stigmergy is termed as an interaction through the environment i.e. two individuals interact indirectly when one of them modifies the environment and the other responds to the new environment at a later time. This is called stigmergy. Ant uses pheromones for stigmergy.

There is a notion of autocatalytic behavior which is defined as the more ants follow a trail, the more attractive that trail becomes for being followed. The process is thus characterized by a positive feedback loop, where the probability of a discrete path choice increases with the number of times the same path was chosen before. This is why ACO algorithms are also called as autocatalytic positive feedback algorithms.

If the query q_i is issued by 12 different users then 12 numbers of ants are assumed to find related suggestions. Essentially ACO implementation takes number of users, queries issued by different users and the terms present in documents clicked by the users. Each ant is a simple agent with the following characteristics:

— It chooses the query term to go to with a transition probability that is a function of the similarity and of the amount of trail present on the connecting edge between terms.
— User navigation over web pages retrieved for a query is treated as ant movement over graph.
— To force the user to make legal transition, navigation to already visited terms are not allowed until a tour is complete.
— When the user completes a tour, a substance called trail or trace or pheromone is laid on each edge (i,j) visited.

In a Graph $G(N,E)$ where N is the number of terms (nodes) and E is the edges and d_{ij} is the edge weight i.e. similarity between i and j. Ant moves from a node i to the next j with the transition probability defined using Eq. (1).

Each edge is associated a static value based on the similarity weight $\eta_{ij} = 1/d_{ij}$. Each edge of the graph is augmented with a trace τ_{ij} deposited by ants (users). Initially it is 0. Trace is dynamic and it is learned at run-time. Each ant tries to produce a complete tour using the probability depending on η_{ij} and τ_{ij} to choose the next term node for possible suggestion.

$$p_{ij}^k(t) = \begin{cases} \dfrac{[\tau_{ij}(t)]^\alpha [\eta_{ij}]^\beta}{\sum\limits_{k \in allowed_k}[\tau_{ik}(t)]^\alpha [\eta_{ik}]^\beta} & \text{if } k \in allowed_k \\ 0 & \text{otherwise} \end{cases} \tag{1}$$

The similarity between any two node in the graph $\tau_{ij}(t)$ denote the intensity of trail on edge (i,j) at time t. Here, the pheromone levels are updated on search session basis in order to preserve the significance of transitions over the graph by users. The edge between q_i and q_j is updated with pheromone level τ_{ij}. For each edge, following trail deposition by all ants i.e. search users, the trail value is updated using the Eq. (2).

$$\tau_{ij}(t+1) = p \times \tau_{ij}(t) + \Delta\tau_{ij} \tag{2}$$

$$\Delta\tau_{ij} = \sum_{k=1}^{m} \Delta\tau_{ij}$$

where p is the rate of trail decay per time interval i.e. pheromone evaporation factor and m is the number of users i.e. ants.

$$\Delta\tau_{i,j}^k = \begin{cases} \dfrac{C}{L_k} & \text{if } (i,j) \in \text{High Similarity} \\ 0 & \text{otherwise} \end{cases}$$

The trail intensity is updated subsequently upon the completion of each algorithm cycle. Each ant subsequently deposits trail of quantity $1/L_k$ on every edge (i,j) visited in its individual tour. The sum of all newly deposited trails is denoted by $\Delta\tau_{ij}$. L_k is defined as the length of the ant i.e. the similarity between the terms visited by the user. C is a constant defined as the average similarity weight of all edges in the graph. Two factors that make the probabilistic model are *Visibility,* denoted by η_{ij}, equals the quantity $1/d_{ij}$ and *Trail,* denoted by $\tau_{ij}(t)$. These two factors play an essential role in the probabilistic transition function of the Ant System. The weight of η_{ij} and τ_{ij} factor in the transition function is controlled by the variables α and β, respectively. Significant study has been undertaken by researchers to derive optimal α and β combinations. A high value for α means that trail is very important and therefore ants tend to choose edges chosen by other ants in the past. On the other hand, low values of α make the algorithm very similar to a greedy algorithm.

4   Experimental Evaluation

4.1   Test Dataset Preparation

The dataset used in this work is AOL search log [14] which possesses implicit feedback in the form of click-through data collected by a search engine and it is released for research purpose. The AOL Search Data is a collection of real query log data that is based on real users. The data set consists of 20M web queries collected from 650k users over the period of three months. We prune this data to suit to our research purpose. For evaluating the proposed approach, only the queries issued by at least 10 users were employed and the pre-processed top 20 documents retrieved for that query were used to construct graph. For example, if the query is issued by 10 different users then 10 numbers of ants are assumed to suggest related queries. 270 single and two word queries issued by different users from AOL search log are taken.

4.2   Baselines for Comparison

The proposed ACO based personalized query reformulation approach has been compared with the following state of the art baseline approaches.

Association Rule Based Approach. The idea behind the association rule based search query reformulation technique proposed in [12] is to use search session boundaries and to treat each search session as a transaction in order to derive related queries from the queries submitted within the same transaction.

SimRank Approach. The query-URL bipartite graph is used to calculate the similarities between queries. Then the top-5 similar queries were suggested to users based on the similarities. The intuition is that two queries are similar if they link to a lot of similar URLs. Besides this; two URLs are very similar if they are clicked as a result of several similar queries. Thus, similarities between URLs are computed and then the similarities for queries are computed based on the similarities of URLs. SimRank approach [13] is purely structure dependent and it does not rely on the domain knowledge.

Backward Random Walk Approach. This approach [19] performs backward random walk starting from the query node on the query-URL graph. After the random walks, top ranked queries are presented as the suggestions.

Forward Random Walk Approach. This approach [19] performs forward random walk starting from the query node on the query-URL graph. After the random walks, top ranked queries are presented as the suggestions.

Traditional ACO Based Approach. ACO based learning from query logs proposed in [1] considers a partial directed graph learned from query log and the weights on the edges represent the importance of the association between the nodes. The graph has been used to recommend queries by starting from the initial query node and then traversing the graph edges to identify the related query nodes. The proposed work considers undirected graph representation to construct graph model. In order to enhance the searching, ternary tree representation is employed to build graph model. Graph is constructed at the end of each word in the ternary tree which increases the speed of searching.

4.3   Evaluation Metrics

The evaluation metrics chosen in this work are Mean Reciprocal Rank (MRR). MRR is defined as given a set of queries Q, the set of documents D, the subset A of which are considered as the relevant documents for $q \in Q$ and the set of first documents $R_{D,Q,n}$ of D returned for every query q so that $R_{D,q,n} = \{D_{q,1}, D_{q,2}, ..., D_{q,n}\}$, the MRR is defined using Eq. (3).

$$MRR(Q, D, n) = \frac{\sum_{q \in Q} RR(q, R_{D,Q,n})}{|Q|} \tag{3}$$

where $RR(q, R_{D,Q,n})$ is the Reciprocal Rank (RR) that depends on the position of the first returned snippet from the result list which contains a relevant document, 0 if there is no relevant document is found in the first n documents.

$$RR(q, R_{D,Q,n}) = \begin{cases} \dfrac{1}{i} \exists i \mid i = \min_{i \le j \le n} j \mid D_{q,j} \in A_{D,q} \\ 0 \qquad otherwise \end{cases}$$

4.4   ACO Parameter Setting

The ACO parameters namely, Depth, evaporation factor, and pheromone updating schemes are set in order to evaluate the proposed approach. The depth refers to the number of hop between the nodes in order to recommend queries. Depth is set as 5 i.e. top ranked 5 related queries. If more than 5 queries are generated, then user will be in trouble choosing which suggestion will be appropriate. The evaporation factor was set to 0 and 0.5. The pheromone trial was updated according to the Eq. (2) by assessing all linked nodes from the query node.

4.5   Experimental Results and Evaluation

Since the dataset used in this paper is different from the datasets that the commercial search engines employ, it is difficult to quantitatively evaluate our results with those from the commercial search engines objectively. Hence, the proposed approach is compared with the baseline approaches namely, Association rule based approach [12], SimRank [13], Backward Random Walk (BRW) [19], and Forward Random Walk (FRW) [19], and the traditional ACO based query log analysis [1].

MRR Evaluation. MRR score is computed based on the ACO parameters considered in order to compare the performance of proposed model with baseline system. Table 1 summarizes the performance of proposed approach and baseline approaches. The percentage of MRR improvement over baseline approaches has been shown in Fig. 3. It is observed that the proposed ACO approach outperforms traditional ACO approach [1] with 22% of improvement.

Table 1. MRR obtained for 10 queries

Approaches	MRR
Association rule based approach	0.387
SimRank	0.362
Backward Random Walk (BRW)	0.314
Forward Random Walk (FRW)	0.321
Traditional ACO based approach	0.411
Proposed query reformulation method using ACO	0.527

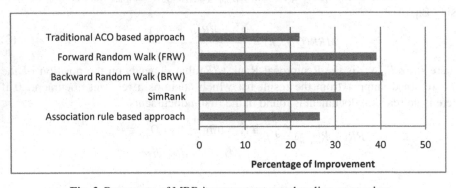

Fig. 3. Percentage of MRR improvement over baseline approaches

From the top 5 personalized suggestions for the initial queries shown in Table 2, it is observed that the query suggestions generated by our method are generally as good as those from commercial search engines. For some queries, our suggestions are even better. In order to compare the proposed method with other approaches, we create a set of 200 queries as the testing queries covering a wide range of topics such as Computers, Arts, Business, Games, Health, and others. Some of the results generated by our ACO based algorithm are shown in Table 2.

Table 2. Top 5 suggestions generated by the proposed method

Testing Queries	Suggestions				
	Top 1	Top 2	Top 3	Top 4	Top 5
Game	Game online	Game stores	Game car	Game theory	Game vui
Graph	Graph theory	Graph function	Graph software	Graph types	Graph plotter
Algorithm	Algorithm analysis	Algorithm ACO	Algorithm programming	Algorithm computer	Algorithm flowchart
Java	Java oracle	Java download	Java updates	Java JVM	Java tutorials
Database	Database relational	Database DBMS	Database examples	Database applications	Database types
Apple	Apple store	Apple iphone	Apple computer	Apple ios	Apple updates
Marathon	Marathon running	Marathon athletic	Marathon fitness	Marathon race	Marathon oil

From the results, we observe that our query reformulation algorithm not only suggests queries which are literally similar to the test queries, but also provides latent semantically related query suggestions. For example, if the test query is 'Algorithm', the proposed approach suggests 'Analysis', 'ACO', 'Program', 'Computer', and 'Flowchart'. All these suggestions have high latent semantic relations to the query 'Algorithm'. All of the results show that our ACO based personalized query suggestion algorithm presents better query suggestions.

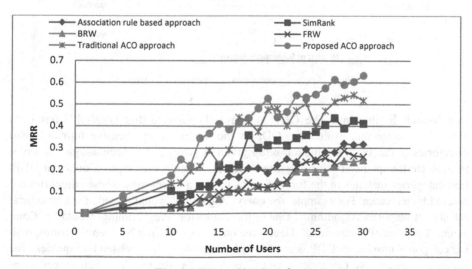

Fig. 4. MRR vs. Number of users

Fig. 4 shows the MRR improvement of the proposed approach. It is inferred that MRR improvement is achieved as the number of search users increased. The proposed approach exploits users navigation over the documents retrieved in order to identify related query suggestions. This approach updates pheromone levels as and when user interacts with search engine. Thereby query suggestions are produced by incorporating users current search interest which is found to be more appropriate and useful.

Manual Evaluation. Users were asked to rate the query suggestion results. The rating score ranges from 0 to 3 (3 - highly relevant, 2 - relevant, 1 - hard to tell, 0 - irrelevant) in order to measure the relevance between the testing queries and the suggested queries in which 0 indicates 'totally irrelevant' while 1 indicates 'entirely relevant'. The average values of evaluation results are shown in Fig. 5. It is observed that, when measuring the results by user experts, the proposed ACO algorithm increases the accuracy for about 21, 26.1, 35.2, 34.7, and 10.6 percent comparing with the Association Rule, SimRank, BRW, FRW, and Traditional ACO algorithm respectively.

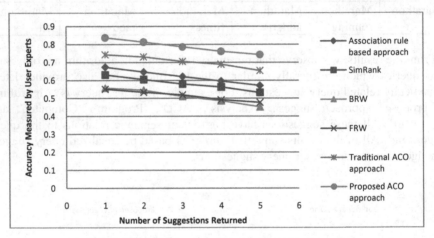

Fig. 5. Accuracy comparisons measured by users

Benchmark Evaluation. The ODP database [http://www.dmoz.org/] [4] has been utilized for automatic evaluation. ODP is the largest comprehensive human edited directories of the Web. Evaluation setup given in [16][22] has been adopted in order to evaluate the quality of the suggested queries. When a user types a query in ODP, find categories matches in the form of paths between directories. These categories are ordered by relevance. For example, the query 'Algorithm' would provide the hierarchical category 'Computers:Algorithms'. One of the results for 'Programming' would be 'Computers:Algorithms:Programming'. Hence, the notion of similarity between the corresponding categories provide by ODP is used in order to measure how related two queries. The similarity between two categories C and C' is defined as the length of their longest common prefix divided by the length of the longest path between C and C'.

$$sim(C,C') = \frac{|LCP(C,C')|}{MAX\{|C|,|C'|\}} \tag{4}$$

where, $|LCP(C,C')|$ is the length of the longest common prefix between C and C', $|C|$ and $|C'|$ are the length of the paths. The length of the two query terms 'Algorithms' and 'Pseudocode' is computed as $2/3=0.667$. Thus, the similarity between two queries is evaluated by measuring the similarity between the most similar categories of the two queries among the top 5 answers provided by ODP.

From Fig. 6, it is observed that the proposed ACO algorithm increases the suggestion accuracy for about 14.4, 20, 31.1, 31.7, and 7.1 percent comparing with the Association Rule, SimRank, BRW, FRW, and Traditional ACO algorithm respectively. This indicates that the proposed ACO based query reformulation algorithm is very effective.

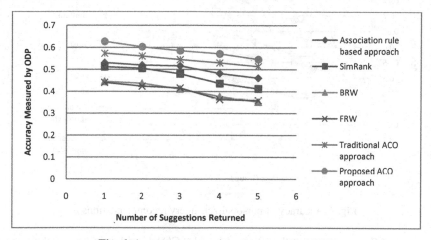

Fig. 6. Accuracy comparisons measured by ODP

The major difference between the proposed ACO based model and the random walk model is that proposed model has the notion of pheromone deposition on edges which defines similarities. Ant is assumed to be traversing from one node to another node based on navigation of users in the past time to time. However, in case of random walk model, the transition from one node to its neighbors is done randomly. Hence, in our model, the suggested query term will preserve more information related to the original query. Thus, the results of proposed approach outperform random walk models.

Evaluating Personalized Query Suggestion. Personalization is becoming more and more important in many applications since it is the best way to understand different information needs from different users [16].

In order to evaluate the quality of our personalized query recommendation approach, we create 10 groups namely $G_1,G_2,...,G_{10}$. G_1 contains users who have submitted only one query. We then randomly select 5 users from the user list for each group. Totally 50 users are considered. For each of these users, ants start the transition with the submitted user query as source. After preparing the results, query suggestions were rated by the expert users. The range of rating score has been defined from 0 to 3 (3 - highly relevant, 2 - relevant, 1 - hard to tell, 0 – irrelevant) in order to measure the relevance between the testing queries and the suggested queries in which 0 indicates 'totally irrelevant' while 1 indicates 'entirely relevant'. The average values of evaluation results for each group are reported in Fig. 7. It is observed that the proposed method generally produces high quality results, and as the number of users increases, the query recommendation quality also increases.

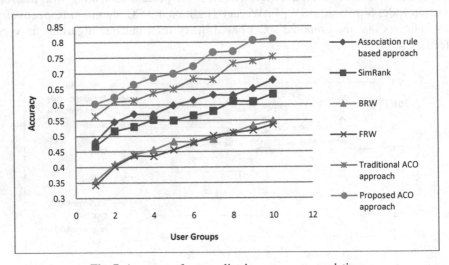

Fig. 7. Accuracy of personalized query recommendations

From Fig. 7, it is observed that the proposed ACO algorithm increases the suggestion accuracy for about 16.56, 21.53, 34.22, 35.53, and 6.88 percent comparing with the Association Rule, SimRank, BRW, FRW, and Traditional ACO algorithm respectively.

5   Conclusion

In this paper, an algorithm for search query suggestion on large scale term graph using ACO principle has been presented. This algorithm is evaluated for personalized query suggestions. The generated query suggestions are semantically related to the initial query. The experimental analysis on large scale web data sources shows the promising future of this approach. The challenge of determining how to suggest the more appropriate query reformulation suggestions for an ambiguous query from a number of possible candidates is addressed in this paper. The ternary search tree data structure is employed for storing query terms and graph is employed to represent terms in the documents. The graph was then used to suggest queries by traversing the graph from the initial query node to discover related query nodes following ACO principle. The experiments were performed on a benchmark dataset AOL query log and the results are shown to be promising.

Acknowledgment. The work presented in this paper is supported and funded by the Department of Science and Technology (DST), Ministry of Science and Technology, Government of India under INSPIRE scheme. Authors wish to extend their thanks to DST.

References

1. Albakour, M.-D., Kruschwitz, U., Nanas, N., Song, D., Fasli, M., De Roeck, A.: Exploring Ant Colony Optimisation for Adaptive Interactive Search. In: Amati, G., Crestani, F. (eds.) ICTIR 2011. LNCS, vol. 6931, pp. 213–224. Springer, Heidelberg (2011)
2. Martins, B., Silva, M.J.: Spelling Correction for Search Engine Queries. In: Vicedo, J.L., Martínez-Barco, P., Muñoz, R., Saiz Noeda, M. (eds.) EsTAL 2004. LNCS (LNAI), vol. 3230, pp. 372–383. Springer, Heidelberg (2004)
3. Manning, C.D., Raghavan, P., Schutze, H.: Introduction to Informa-tion Retrieval. Cambridge University Press (2008)
4. Carpineto, C., Romano, G.: ODP239 dataset (2009), http://credo.fub.it/odp239/
5. Cui, H., Wen, J.-R., Nie, J.-Y., Ma, W.-Y.: Probabilistic query expansion using query logs. In: Proc. 11th Intl. Conf. on World Wide Web, pp. 325–332. ACM (2002)
6. Cui, H., Wen, J.-R., Nie, J.-Y., Ma, W.-Y.: Query expansion by mining user logs. IEEE Trans. Knowledge and Data Engineering 15(4), 829–839 (2003)
7. Dignum, S., Kruschwitz, U., Fasli, M., Kim, Y., Song, D., Cervino, U., De Roeck, A.: Incorporating Seasonality into Search Suggestions Derived from Intranet Query Logs. In: Proc. IEEE/ACM Intl. Conf. Web Intelligence and Intelligent Agent Technology, pp. 425–430 (2010)
8. Marco, D., Maniezzo, V., Colorni, A.: Ant system: optimization by a colony of cooperating agents. IEEE Transactions on Systems, Man, and Cybernetics-Part B 26(1), 29–41 (1996)
9. Dorigo, M., Birattari, M., Stutzle, T.: Ant colony optimization. IEEE Computational Intelligence Magazine 1(4), 28–39 (2006)
10. Efthimiadis, E.N.: Query Expansion. In: Williams Martha, E. (ed.) Annual Review of Information Systems and Technology, vol. 31, pp. 121–187. Information Today (1996)
11. Agichtein, E., Brill, E., Dumais, S.: Improving Web Search Ranking by In-corporating User Behavior Information. In: Proc. 29th ACM SIGIR Intl. Conf. Research and Development in Information Retrieval, pp. 19–26 (2006)
12. Fonseca, B.M., Golgher, P.B., de Moura, E.S., Possas, B., Ziviani, N.: Discovering search engine related queries using association rules. Journal of Web Engineering 2(4), 215–227 (2003)
13. Jeh, G., Widom, J.: Simrank: A Measure of Structural-Context Similarity. In: Proc. 8th ACM SIGKDD Intl. Conf. Knowledge Discovery and Data Mining, pp. 538–543 (2002)
14. Pass, G., Chowdhury, A., Torgeson, C.: A Picture of Search. In: Proc. 1st Intl. Conf. on Scalable Information Systems (2006)
15. Cui, H., Wen, J.-R., Nie, J.-Y., Ma, W.-Y.: Query Expansion by Mining User Logs. IEEE Trans. Knowledge and Data Engineering 15(4), 829–839 (2003)

16. Ma, H., King, I., Lyu, M.R.-T.: Mining Web Graphs for Recommen-dations. IEEE Trans. Knowledge and Data Engineering 24(6), 1051–1064 (2012)
17. Jansen, B.J., Spink, A., Saracevic, T.: Real life, real users, and real needs: a study and analysis of user queries on the web. Information Processing and Management 36(2), 207–227 (2000)
18. Kelly, D., Gyllstrom, K., Bailey, E.W.: A comparison of query and term suggestion features for interactive searching. In: Proc. SIGIR, pp. 371–378. ACM (2009)
19. Craswell, N., Szummer, M.: Random Walks on the Click Graph. In: Proc. 30th Annual Intl. ACM SIGIR Conf. Research and Development in Information Retrieval, pp. 239–246 (2007)
20. Chirita, P.-A., Firan, C.S., Nejdl, W.: Personalized Query Ex-pansion for the Web. In: Proc. 30th Intl. ACM SIGIR Conf. Research and Development in Information Retrieval, pp. 7–14 (2007)
21. Kraft, R., Zien, J.: Mining Anchor Text for Query Refinement. In: Proc 13th ACM Intl. Conf. World Wide Web, pp. 666–674 (2004)
22. Baeza-Yates, R., Tiberi, A.: Extracting Semantic Relations from Query Logs. In: Proc. 13th ACM SIGKDD Intl. Conf. Knowledge Discovery and Data Mining, pp. 76–85 (2007)
23. Baeza-Yates, R., Ribeiro-Neto, B.: Modern Information Retrieval. Addison-Wesley (1999)
24. Jones, R., Rey, B., Madani, O., Greiner, W.: Generating Query Substi-tutions. In: Proc. 15th Intl. ACM Conf. World Wide Web, pp. 387–396 (2006)
25. Wang, X., Zhai, C.: Learn from Web Search Logs to Organize Search Results. In: Proc. 30th ACM SIGIR Intl. Conf. Research and Development in Information Retrieval, pp. 87–94 (2007)
26. Yin, Z., Shokouhi, M., Craswell, N.: Query Expansion Using External Evidence. In: Boughanem, M., Berrut, C., Mothe, J., Soule-Dupuy, C. (eds.) ECIR 2009. LNCS, vol. 5478, pp. 362–374. Springer, Heidelberg (2009)

SemEnAl: Using Semantics for Accelerating Environmental Analytical Model Discovery

Kalapriya Kannan, Biplav Srivastava, Rosario U.-Sosa,
Robert J. Schloss, and Xiao Liu

{kalapriya,sbiplav}@in.ibm.com,
{rosariou,rschloss}@us.ibm.com,
liuxiao@sg.ibm.com

Abstract. Web as a platform to integrate applications, encapsulated as web services and composed using semantic technologies, is well established. However, in many domains, there are far more applications whose description or implementation are available than those using service-enabled access. For example, in the environmental domain, analytical models (computational functions) are crucial to analyze collected sensor data, extrapolate them for uncovered but interesting settings of interest, and gain insights for action. However, using a particular analytical model may be appropriate under very specific conditions - terrain of region, weather conditions, pollutants, types of pollutant sources, data sampling rate, etc. Thus, finding a relevant analytical model for a given setting is of particular interest to environmental agencies around the world. But it is also a complex activity since there are hundreds of analytical models with numerous constraints. In this paper, we present *SemEnAl*, an approach to use semantic annotations derived from a domain ontology to address this discovery problem and demonstrate its effectiveness. Our experiments with search system for analytical models indicates that using *SemEnAl* can provide high precision, to an extent of about 100% relevant models for selected queries. The paper thus pushes the reach of semantics to new domains.

1   Introduction

Environmental issues such as Air Pollution and Quality (APQ) are a prominent concern for citizens and cities. To monitor them and take timely action, environmental engineers collect selected data from field sensors at a limited number of locations, extrapolate them for uncovered – but interesting – settings of interest (e.g., additional locations, future times), analyze the data using sophisticated algorithms implemented with diverse programming technologies or computed manually, and initiate actions for response (e.g., issue alerts for citizens to stay away from concentrations of pollutants during certain times).

The algorithms to extrapolate and analyze data are also known as analytical models (AMs). An AM may be appropriate under very specific conditions -

S. Srinivasa and S. Mehta (Eds.): BDA 2014, LNCS 8883, pp. 95–113, 2014.
© Springer International Publishing Switzerland 2014

terrain type of the region region, specific weather conditions, specific classes of pollutants, types of pollutant sources, data sampling rate, etc. Thus, finding a relevant analytical model for a given setting is of particular interest to environmental agencies around the world. But it is also a complex activity since there are hundreds of analytical models with numerous constraints; running an inappropriate model wastes IT resources and can lead to harmful inaccurate actions; failing to run a predictive model robs officials of meaningful insight. This paper describes our framework to accelerate discovery of relevant AMs for settings of interest, and if their implementation is available, enable their seamless execution.

For this problem, if applications were available as services, the web has been shown to be a promising platform to integrate applications using semantics-enabled composition techniques[2]. However, in many domains, there are far more applications (algorithms) whose description or implementation is available than those with their REST-style service-enabled access. Environment happens to be one such domain. Here, it is practically infeasible for the user to implement every algorithm or service-enable every implementation unless they are relevant to his setting. So, just discovering the details or implementations of the right applications for a specific setting, would help them prioritize their integration effort.

Our analysis of about 150 AM's in the space of APQ from diverse sources [13] [11] indicates that the problem of identifying suitable models for specific criteria goes beyond mere availability of data for executing them. The reason for this are several. **Firstly**, AMs are derived from fundamental equation of dynamics such as heat transfer, mass, weather conditions, regional topographic characteristics etc. As a result, the final equations -which are the core of the AM's once obtained for one case may not be used/useful in for another case. Thus for a query of nature 'obtain the AM that use wind data' returns about 70% models, but each of them operates with different wind setting, such as specific speed, direction etc., This reduces the relevant models applicable for a given data set. **Secondly**, number of input data are numerous (we have identified about about 200 distinct input data all together for 150 models). This along with the conditions of applicability of a model for a suitable scenario explodes. For instance, one of the commonly required input data is the pollutant concentration level, but the model is specific to the nature of the pollutant source. Thus, the major challenge beyond developing these models is identifying the suitable model based on finer details of developing the individual models. We use the term, Operating Conditions (OC), to refer to those conditions the determines the suitability of model at a specific location or time. **Thirdly**, the nature of the queries an environmental agencies is typically interested is time-space 'forecasting'. The AMs directly do not relate to the queries (specific to location or time) through the input or other OCs. Thus, determining suitable models for a space or time requires capabilities beyond mere data analysis to co-relate information about locations/time in a way that it can be equated or related to the input data or OC's. Such capabilities have so far been limited to domain matter experts who have through understanding of the models (available through the technical specifications). While the capability requirement remains same for any execution of the AM, the human intervention has a negative effect on the re-usability of the knowledge in another setting.

In an attempt to simplify the identification and utilization of suitable models, we propose $SemEnAL$ that leverages the ontology world to capture the semantics of the OC. These OC's are modeled in the form of constraints to invoke the models. We achieve this goal in two phases. In the first phase, we build the semantic network of the key concepts in the air pollution domain. We further model the analytical models available in this domain using the traditional conventions of semantic web services. The semantics of OC are captured in the form of constraints to these models. In the second phase, we integrate the relationships between the concepts of the air pollution domain to the requirements of the analytical models. Such integration would provide automatic siphoning of the requirements of the Analytical models that are directly utilized from the different key concepts related to pollutants and its nature such as pollutant sources, effects and categories of both pollutant and pollutant sources. We further believe that such integration would help us identify suitable and accurate analytic model with higher precision leading to narrowing the search results.

Our contributions in the paper can be summarized as follows:

1. We bridge the key concepts of the domain 'Air Pollution' to the analytical models available in the domain such that the data flow (for inputs) and preconditions related to Analytical models can be directly obtained from the domain concepts.
2. Semantically capture the requirements of the Analytical models for both input (its ranges) and the preconditions.
3. Shows such annotation and bridging will narrow the search scope and the results obtained for the determining the suitability of the analytical models for specific regions or time.

The rest of the paper is organized as follows. We elicit the problem challenges with a running example in Section 2. Our solution overview and details is presented in Section 3. Our experimental platform to validate our hypothesis is presented in Section 4. We conclude our paper in Section 6.

2   Running Examples: Selected AM*s* from Domain APQ

We begin with detailed analysis of a simple AM called Gaussian Plume Model (GPM) and also show the higher orders of complexity with complex models such as CALPUFF.

GPM has received attention due to the gaining importance of emissions from stationary points sources. Such sources are alternatively referred to as elevated stack specific point emissions. Figure 1 provides an illustration of the model. From the diagram it can be seen that plume (column) has several parameters h_s - Height of the stack, H is the effective stack height which equals the sum of the physical stack height and the plume rise, u-wind speed at effective height,

σ_y and σ_z are respectively the standard deviation of pollutant concentration in the cross-wind and vertical directions and are determined by Pasquill-Gifford stability categories [20], x, y - denotes downwind and crosswind distances from the stack to the site of interest, z is the height of the site in the vertical direction. Along with the above input data, the mass emission rate $(\mu g/s)$ is utilized to compute the concentration levels of the pollutants. It should be noted that in the computation of σ_y and σ_z using Pasquill-Gifford requires Solar radiation during the day and cloud cover at night are required to determine the stability category. Readers are advised refer to [7] to get details of the GPM that utilizes the above parameters to compute the background Pollutant Concentration (PC) level.

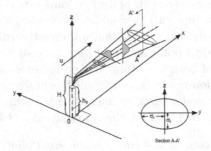

Fig. 1. Illustration of the Gaussian Plume Model [9]

To summarize, GPM is based on a few key assumptions, such as the plume spread has a Gaussian distribution, the emission rate is constant and continuous, uniform wind speed and direction, total reflection of the plume takes place at the earth's surface, i.e., conservation of mass, and so on. The output of the GPM is a singleton value either a real value indicating the concentration level of the pollutant or '-1' indicating the non-availability of the concentration level due to missing input value or zero indicating the absence of pollutant at specific time. Table 1 summarizes the input and output for executing the *GPM*.

Table 1. IO of GPM

Input	Output
Height of stack, Effective Height of Stack, Radius/diameter of the stack, wind speed at effective height, atmospheric stability parameters, standard deviation of pollutant concentration based on wind parameters, downwind and crosswind distances,height of the site	Real Value - indicating concentration level
	Not Available - indicating a missing mandatory input
	Zero - Either truncation of a insignificant value or no observed pollutant

One of the common requirements of the agencies concerned with pollution control is to identify the models that are specific to a region and at specific time.

Let us consider a query of the form *'Is this model suitable for a location whose zip-code is* 110024*?'*. Given the input and output of GPM in Table 1, the resolution to this query is impractical as the location code zip-code is not directly considered as an input to the GPM. It is left to the experts to conduct in-depth study of the user guide to identify whether the characteristics defined by the region are used in the Analytical models. For instance is there an point source pollutant emitting source available in this location, such that this model can be applied (or)location of a Gas Turbine in this location makes this model a potential candidate to be executed in this environment. This information typically an *OC* requires information from external domains (such as city infrastructure etc.,), nature of pollutant (APQ domain) to resolve the query. It requires experts with high level of domain knowledge and skills to understand the various operating conditions thus making even the simplest of model extremely difficult for easy adoption and utilization. Since the operating conditions of the model are not readily available in the form of the input data, the only way to effectively capture them is to model them in the form of constraints.

Our similar analysis of another complex *AM* called CALPUFF [15] shows similar concern. Again given the query above, the input and the output provided by this model is insufficient to resolve the query. This sets motivation for our problem to capture the hidden information about the *AM*'s OC in a structured manner through semantic networks. We believe that such enablement will provide scope for deep inferencing, thereby such queries can be supported through multiple output chaining.

3   Solution Overview

Our aim is to allow concepts or semantic information from the domain of *APQ* to be integrated with the semantic concepts of *AM* and show that through such integration both information flow and inferencing can be enabled. Figure 2 illustrates the different phases that are present in our approach.

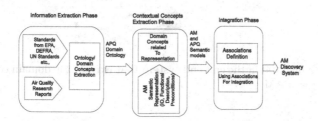

Fig. 2. Phases in Extraction of Air Quality Domain Concepts

In the Information Extraction Phase (IEP) we identify the concepts in the domain of *APQ*. Once the concepts are identified the concepts phase through the Contextual Concepts Extraction Phase (CCEP) in which concepts that are required in the context of *AM*'s are identified. Once the relevant domain concepts are identified the AM's are integrated with them through *Associations* in the Integration Phase (IP). These associations are defined one time and are reusable for all *AM*'s.

Figure 3 gives the process overview of the end-to-end system. *IEP* phase requires extraction of concepts related to APQ from several input documents. Given that there are existing technologies that can identify the key concepts from the domain from unstructured text, we did not need to re-invent. As an illustration of the state-of-the-art, we work with IBM SPSS Text Analytics [14] which is a commercial tool that works with PDF, Word and other popular document formats. SPSS performs unsupervised extraction and can generate concept hierarchies. It first separates textual data from other content in the document and then identifies candidate concepts which can be made up of both single or multiple words. Then, it resolves synonyms and organizes concepts in a canonical form. The extracted concepts can be used to build a hierarchy. For this, SPSS uses a combination of linguistics and statistical methods to extract relationships among concepts. Further details on the techniques are available in its help documentation [14].

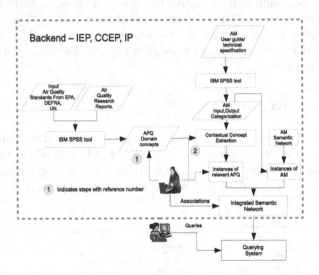

Fig. 3. Processes involved in the end-end system that extract the domain concepts and integrates them with AM

The concepts extracted are fed into the Contextual Content Extraction that identifies the relevant *APQ* concepts for the *AM*. For this purpose, Contextual Content Extraction utilizes Input,Output and other information extracted from the technical specifications of the *AM* again with the help of SPSS tool. The requirements for the executing *AM* is captured in the form of AM semantic network. We borrow concepts from the world of semantic web services [17] to represent the *AM* in the form of *IOPE − Input, Output, Precondition, Effects*.

For all *AM* (sample of about 40) we identify all the inputs, outputs, preconditions and effects and create instances of them. Similarly, for all the relevant
APQ identified we create instances and is used in the Integration Phase for creating the Integrated Semantic Network. This network is presented through the
querying system to the consumers who can then query for suitable sub-models
given different criteria.

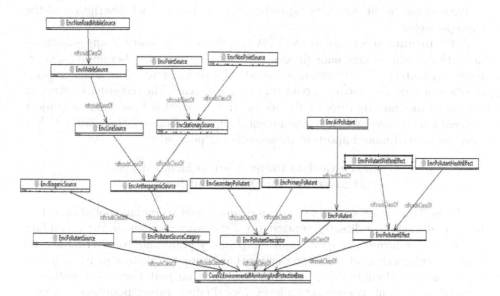

Fig. 4. Pollutant, Pollutant Source categories and Effects

3.1 IEP Phase

In order to define an ontology, we need to define (i) classes, (ii) data properties
(data fields) and (iii) object properties (or relations). Classes define the key concepts in the domain under consideration and properties are used to describe the
characteristics of the concepts. On a relatively less explored domain as *APQ* for
ontologies, one method of determining concepts is to utilize the wide document
base that provide information about the domain. We had collected documents
related to Air Pollution specifically the standards, the measurement techniques,
analytical functions and other air control and regulation from several leading environmental agencies (about 6 covering the most mature and established agencies
[EPA of US, DEFRA of UK [23], WHO of UN [18], NEA of Singapore[3], Danish
EPA, European Union Commission]) and their portals. Using the IBM SPSS,
major concepts and their categories were identified from these sources. The tool
retrieves about 146 sub-concepts all categorized into major categories. Further
analysis through human intervention (as observed in Step 1 of Figure 3) resulted
in pruning 8 important concepts capturing the pollutant related information in
APQ. Table 2 presents eight concepts identified from the documents.

Table 2. Concepts from Air Quality Domain

Pollutant, Pollutant Source, Effect, Area/Region, Measurements, Pollutant Indicators and KPI, Analytical Functions, Pollutant Monitoring Stations.

We enumerate the concepts captured by providing a brief description of the concepts below.

A **Pollutants** according to the EPA [11] business glossary is any substance in air that could, in high enough concentrations harm man, other animals, vegetation, or material. Pollutants may include almost any natural or artificial composition or airborne matter capable of being airborne. The hazardous nature of the pollutants and its direct influence on the human life have made a subject of great concern over all the environmental agencies. Some of them listed below were considered more important presence being pervasive.

Ozone, Particulate Matter, Carbon Monoxide, Nitrogen Oxides, Sulphur Dioxide and Lead

The pollutants are of two categories primary and secondary pollutants. Primary pollutants are directly emitted from the sources whereas the secondary pollutants are formed by reaction of the primary pollutants to external factors such as sunlight/rain etc. The properties that define the pollutants are physical type, name, availability type, pollutant pathways, air pollution index, pollutant ranking assessment, concentration level, Moral effect, target pollutant, toxicity, indicators, health effects, welfare effects, environmental effects. These properties capture the attributes of the pollutants and only a subset (as will be established later) will be directly required for executing the *AM*.

The pollutant source is another important concept in the air quality domain for its contribution to the pollutant levels. The pollutant source has different subcategories. A pollutant at the outermost classification is either anthropogenic source (man made sources) or biogenic source (natural sources like volcanoes etc.,). Anthropogenic sources are of particular importance due to their expanding base and are further classified into stationary and mobile sources. A stationary source in turn is either point source(when a single source can be attributed to the emission) or area source (when the pollutant is part of the area or generated from an area). Mobile sources can be either those originating from the road or otherwise non-road mobile sources such as construction facility/gaseous powered lane.

The third important concept is the effects that are caused/ induced by the pollutants. Depending on the nature of the object that are affected by pollutants, the effects are either classified as health effects or welfare effects. Health effects depending on the level and nature of impact can be classified chronic or acute. Figure 4 shows the taxonomic classification and ontological concept captured around the Pollutants, their sources and their effects.

Pollutants are measured in two different contexts. One, when a source emits pollutant and another when pollutant are measured by a measuring station (which is not necessarily a pollutant source). The former is required for source attrition whereas the latter has more passive role. We introduce two concepts called "Measurable Property (MP)" and "Measured Observation (MO)" to seperate the two concepts. MP is used to refer to emission from the pollutant source. For instance a industrial source emits pollutants and therefore the pollutant should be associated using MP to the source. MO refers to an output/observation provided by a measuring instrument. For instance, measuring station employ (not necessarily the sources of pollutants) to observe the different concentration levels.

Indicators and KPIs, define acceptable and optimal levels of pollution -or any other dimension under consideration- for the city. Each pollutant has a KPI defined which has units, measurement duration and value associated with it.

To the best of our knowledge the work presented in this paper is the first in the domain of APQ to capture exhaustive key concepts and indepth classification of the high level concepts. Our structured semantic models can be directly be consumed for practical applications such as model discovery, validation and execution platform etc.,

3.2 Contextual Concepts Extraction Phase

In this phase we define Semantic Model's (SM) to capture the requirements (input, output and OC) of AMs. This SM would form the fundamental premise of identifying the requirements (input data, specific conditions if any on the data etc.,) for realizing (i.e., ready for execution) the AM. We use a simple IOPE (input, output, preconditions (which define the OC) and Effects) representation used earlier in several similar context as such capturing the semantics of Web Services as in WSDL-S [4]. Thus we are interested in capturing the input, output and other OC's that determine the suitability of realizing a model.

Both our semantic representation of analytic modules and WSDL-S address the same problem: that of discover (analytic) services based on how their inputs and outputs map to a semantic model of the domain. From this point of view, our Semantic Model can be integrated in a WSDL-S directory of services. Furthermore, this model-based approach enables an integrated, rich, semantically navigable catalog of analytics for Environment Monitoring and related domains, like Weather, Geospatial, Demographics, etc.

Our observation from manual inspection of 40 AM indicates that all concepts presented in Table 2 are not required for input output of AM's. In order to identify the required concepts, we use collection of documents (user guide and technical specification) related to these 40 AM's and using IBM SPSS we extract input/output/other OC. Table 3 provides the different inputs and their categories. For sake of brevity we do not enumerate the entire list of input/preconditions identified, but present the sample list of input and categories which they belong to.

From Table 2, using Table 3, we identify that the classes required from APQ in the context of the AM are Pollutant, Pollutant Source, Pollutant Source

Table 3. Input,Output,Preconditions Categorization

Category	Input/conditions
Meteorological information	wind speed, wind direction, Minimum and maximum ambient air temperatures,Anemometer height,Surface characteristics, humidity at specific times
Source inputs	Point/volume/area/line, stacks details, area, grid size, scale, elevation, downwash, concentration levels
Landform details	water, deciduous forest, coniferous forest, swamp, cultivatable land, grassland, urban, rural, desert, canyons, terrains, building effects, road geometrics, number of intersections
Receptor options	Receptor location, height, scale, width, diameter
Emission data	reactive decay, reactivity class, dry/wet deposition

Categories, Measuring Stations (also referred to as Receptors in Table 3). This process required human expertise to identify the concepts that might have directly relationships to any input/ output/ preconditions of the AM's (as shown in Step 2 of Figure 3).

Concepts Related to Air Quality from other Domains. This section discusses a selected set of concepts (classes) from other domains that influence Air Quality at a specific location. We borrow relevant concepts from other domains and build relationships with Air Quality domain concepts. **Landforms:** Although area/region is not a concept of Air Quality, pollutants are associated to a specific region/area and often this information has been captured to identify the source of the pollutants. For instance a industrial source may be located a specific region/area and can be the source of pollutants. Thus area should be associated to the pollution stations. **Weather or Meteorological Conditions:** Air quality is largely influenced by the meteorological conditions at a specific time and location. Information such as the humidity, wind speed, wind direction, surface temperature, atmospheric stability all factors that affect the pollution concentration levels. **Traffic Information:** Traffic sources have been identified as one of the major source of pollutants. Information related to traffic volume, traffic pattern are all part of traffic domain. **Road Geometrics:** Air quality is also dependent on the number of intersections, the length of the road (due to estimation of the level of traffic) etc.,

We believe that a similar exercise preformed in Section 3.1 and Section 3.2 will allow identification of concepts from these domains that might directly relate to IOPE of AMs.

Table 4. Associations integrating the concepts of APQ and *AM*

PropertyName	Domain	Range
pollutionObservation AssociatedToSource	MeasuredObservation	MeasuringSource, PollutantSource
pollutantAssociatedTo PollutionObserva-tion	Pollutant	MeasuredObservation
pollutantAssociatedTo	Pollutant	PollutantSource
analyticsHas Pre-condition	AnalyticModule	AnalyticsConstraint
inputToAnalytics	AnalyticModule	AnalyticsInput
outputToAnalytics	AnalyticModule	AnalyticsOutput

3.3   Integration Phase

In this phase we create instances of *AM* for all the models identified and all relevant concepts of *APQ* and integrate them through *associations* such that the inputs and preconditions are associated to the concepts of the *APQ*. Table 4 and Table 5 provides a listing of the associations and relationship respectively that relates the concepts of *APQ* and input or precondition of *AM*.

Table 5. Relationship between concepts in *AM* and concepts in 'APQ'

Class	subclassOf
MeasurableObservation	AnalyticsInput
MeasurableProperty	AnalyticsOutput
AnalyticsCondition	ConditionByClass
AnalyticsCondition	ConditionByBoolean

Figure 5 shows the integration of *AM* and *APQ* concepts by using the running example of *GPM* (explained in Section 2). The figure consists of set of domain concepts, instances of the domain concepts and the association between them. These integrated semantic networks across different domain provides scope of higher level inferences across domains which are otherwise far fetching. Such integration will support queries referred in Section 2 where models can be iden-tified based on the location and inferences based on the availability of pollutant sources can be performed. Thus this integrated semantic network between the domain concepts and *AM* will open new search criteria addressing the practical requirements of finding and discovering suitable AMs.

4   Assessment and Evaluation

4.1   Experimental Setup and Data Sets

Our data set consists of exhaustive list of the AMs for measuring the Air Quality. The major source of Analytics models along with their functional descriptions

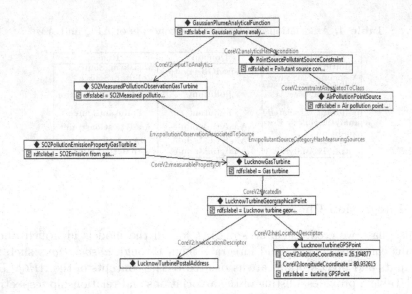

Fig. 5. Gaussian Plume Analytical Function with sample Preconditions and Input

and user guides have been US EPA repository [11] and UK [23], while a few were collected from other Environmental agencies and research reports [13]. We have collected around 40 dispersion functions in the usable form (including their execution requirements). Our hypothesis is based on the fact that the non-trivial enhancement of analytics function discovery using semantic information will narrow the search results for AM discovery and reduce the ambiguity of identifying suitable models for different query options. Thus we build a query system that consists of the following platform:

1. Base Line Platform (*BLP*): Consisting of AM along with their input and output.
2. Functional Description Platform (*FDP*): Comprises of *BLP* along with the tags that describe the functional aspects of the the AMs. Examples include, the category of the AM (Gaussian Plume, Gaussian Puff etc.,), applicability of AMs from the distance of pollutant source and so on.
3. Semantic Enhancement Platform (*SEP*): Comprises *FDP* along with enhancements using semantic information such as nature of the pollutant source, the weather conditions, the different land characteristics that are processed and modeled within the AMs.

Input parameters/variables and output for *BLP* of the AMs are exhaustively obtained by carefully analyzing the execution platform along with the sample input data for executing the AM. When no executable was available we fall back on the technical documents describing the input and output for the analytical models. For input, we collect the the different parameters, their format and their value ranges . For instance the input to a model can be building height, but the model is applicable only for buildings with certain height range. While the inputs were readily available for the model by analyzing the sample input data, the ranges were obtained through deep manual inspection of their user guide specification. Similarly we collect the output information specific to concentration levels, pollutant list, unit of measurement and the duration for which the measurement was performed. The models are typically designed to follow and provide output according to defined KPI's for air quality standards. For instance hourly/annual/8-hrs/24-hrs/3-hrs mean as output frequency is quiet common.

We noticed that the suitability of the model for specific conditions goes beyond just the input and the output data that we had collected for individual models. We utilized the functional description of the AMs that are available as part of Technical specification of the models. We set out to identify the function description of the analytical models (FDP). The functional description consists of description of the AM, the purpose, processes it models, the scale, resolution etc., that the function applies too. Although the technical description can provide the functional description, not all concepts in the technical specifications are important for function selectivity. We have used the technical expertise of several air pollution modeling experts to filter the appropriate category of information that is part of functional description; these act as filters for selecting the appropriate functions. Table 6 provides category and actual information collected from function description.

Table 6. Function Description and their Values

Category	Feasible values
Model Scale	Local, regional and global
Grid Size	sq kms
Resolution size	height and width
Pollutants	CO, NO, NOx, NO2, CO2, SOx, O3, Gases and Particulate Matter
Model Category	Box, Gaussian Plume, Gaussian Puff, Lagrangian, Eulerian, Computational Fluid Dynamics
Atmospheric Stability	Pasquill,Boundary Layer Scaling, Turner

SEP are also identified from the technical specification, but SEP's lead to direct inferencing for identifying suitable models based on the different characteristics of the actual location where the air quality is intended to be measured. Unlike functional descriptors that depict the analytic functions static functional properties the SEP's capture the dynamic aspects that are available in the environment but affect the applicability of a model. Table 7 presents the different constraints captured for functional models.

Information presented in Table 7 forms the most important inferencing (and therefore a filtering) criteria for identifying the suitable models. For instance, an AM is suitable to be executed in a specific location based on the nature of the land form characteristics that constitute the basic contour of the location. In the similar manner, the proximity of the type of the pollutant source such as line/area/volume/point is an major influencing parameter for the suitable AM that can used to estimate the air quality at specific locations.

We build a semantic network for all of the 40 AMs with their input/output (I/O), functional description and their SEP's modeled as preconditions. SEP's unless and otherwise they are satisfied, the models will not be suitable to be executed in specific environment. Listing below provides the AMs considered for our experimentation (majority of them listed under EPA [11] as preferred model).

> AURORA, CPB, CALINE4, HIWAY2, CAR-FMI, AEROPOL, ADMS, GRAL, GATOR, OSPM, STAR-CD, ARIA LOCAL, PBM, CALPUFF, SCREEN3, TAPM, AERMOD, SPRAY, MISKAM, MICRO-CALGRID, BLP, CALINE3, CTDMPLUS, OCD, AERSCREEN, ADMS4, AFTOX, CAMX, CAP88, CMAQ, CAL3QHCR, CTScreen, DENSE GAS DISPERSION MODEL, ISC3, MMFRST, MOTOR VEHICLE EMISSION SIMULATOR, ASPEN, HYROAD, ISC PRIME, OBODM

Table 7. Constraint captured for various Air Dispersion Models

Constraint	Values
Pollutant Source Category	line, volume, area and point
Land form characteristics	Building Wakeup effect, Terrain, Street Canyon, Topography, Intersections,Plume Rise,Aerosols
Turbulence processed within the AM	Turbulence of Ambiance Air(AMB), Vehicle Induced Turbulence (VIT)

4.2   Evaluation

In accordance with our hypothesis, our objective is to measure two factors (a) that AM discovery through semantic information narrows the search results and provides relevant models for queries (b) that semantic inferencing on SEP's opens new search criteria that has practical impact that are otherwise not feasible with the input/output or functional description.

AM Discovery through Semantic Enhancement. We identify a set of queries commonly used in practical scenarios by environmental agency members to determine air pollutant levels. In order to collect the queries we had interacted with both the air pollution domain experts and novice users of the AM's from environmental agencies. We allowed them to submit their queries through a simple interaction interface. Listing below provides the set of queries that were collated and used to evaluate our system.

1. Q1: Models based on the pollutants. Example Query- Find the models that can be used to determine CO_2 levels.
2. Q2: Models based on the pollutant source type. Example Query- Find the models that can be used to determine air pollution due to point sources.
3. Q3: Models based on the weather conditions. Example Query- Find the models that are applicable when the wind is strong and the wind direction is North.
4. Q4: Models based on the wind turbulence. Example Query- Find the models that are suitable to determine air quality when the turbulence in ambient air is high.
5. Q5:Models based on the landforms/topography. Example Query- Find the models that can be used when street canyons are available.

We propose to use precision and recall metric to identify the set of models suitable given a input query. Precision is measured by the fraction of retrieved AMs that are relevant. Let N_{total} represent the total number of the AMs used for evaluating the system, N_r be the relevant models available in the original set of AMs for the intended purpose, N_{ret} be the number of AM retrieved by the query, N_{rrel} is the number of relevant models retrieved by the query then Equation 1 present the equation for computing the precision.

$$P = \frac{N_{rrel}}{N_{ret}} \tag{1}$$

Recall is measured by the number of relevant AMs retrieved given the total number of actual relevant AM available in the original set of AMs. Equation 2 presents the equation for measuring the recall given a query output.

$$R = \frac{N_{rrel}}{N_r} \tag{2}$$

Table 8 presents the number of suitable analytical models for different search criteria. We use the results presented in this to compute the precision and the recall. Equations 1 and Equation 2 were used to compute precision and recall respectively. It can be observed that the precision using the SEP (semantic information) is as high as 1 (100%) not only in cases where the queries are predominantly depended on the IO but more importantly in cases where IO does not provide any matches ($Q1, Q2, Q3, Q4, Q5$). On the other hand, using only IO yields poor precision (zero) even on some of the most practical and fundamental search queries $Q2, Q3, Q4, Q5$ such as identifying models based on the pollutant name ($Q2$). In the similar manner the recall while using the semantic enhancement is as high as 1 for identifying relevant models. Higher recall indicates the false positives returned are low with semantic information. This establishes the superiority of using the semantic enhancement for AM discovery in the area of environmental domain specifically with air quality.

Table 8. Query results for determining suitable models [P=Precision, R= Recall]

Query No	Purpose specifics and Description	Querying Platform			Number of relevant models in the original set.
		BLP	FDP	SEP	N_r
Q1	*AM* that takes pollutants as input	8(P=R=1)	8 (P=R=1)	8 (P=R=1)	8
	AM that take pollutant background concentration	13(P=R=1)	13(P=R=1)	13(P=R=1)	13
	AM that are used for computing CO content in air	0(P=R=0)	38 (P=R=1)	38(P=R=1)	38
	AM that are used for NO content in air	0(P=R=0)	38 (P=R=1)	38 (P=R=1)	38
Q2	*AM* that are applicable for pollutant source being point source	0(P=R=0)	0 (P=R=0)	26(P=R=1)	26
	AM that are applicable for pollutant source being volume source	0(P=R=0)	0 (P=R=0)	15(P=R=1)	15
	AM that are applicable for pollutant source being area source	0(P=R=0)	0(P=R=0)	18 (P=R=1)	18
	AM that are applicable for pollutant source being line source	0(P=R=0)	0 (P=R=0)	23 (P=R=0.7)	18
Q3	*AM* that consider wind speed	26(P=R=1)	26 (P=R=1)	26 (P=R=1)	26
	AM that consider wind direction	26(P=R=1)	26(P=R=1)	26 (P=R=1)	26
Q4	*AM* that consider current air turbulence	0(P=R=0)	0 (P=R=0)	21 (P=R=1)	21
	AM that consider vehicle movement/vehicle induced turbulence	0(P=R=0)	0 (P=R=0)	12 (P=R=1)	12
Q5	*AM* that models building wake effect	0(P=R=0)	0 (P=R=0)	13 (P=R=1)	13
	AM that models terrains and complex topography	0(P=R=0)	0(P=R=0)	13(P=R=1)	13
	AM that models chemistry of pollutants	0(P=R=0)	0(P=R=0)	8 (P=R=1)	8
	AM that models street canyons	0(P=R=0)	0(P=R=0)	9 (P=R=1)	9

Semantic Inferencing Leading to New Search Criteria. One of the most important problem that we identified from our joint exercise with some of the environmental agencies is identification of the AM based on location and future time. Listing below provides the set of queries that end users of the *AM* would use to identify suitability.

> Q6: Identify models that can be executed at a
> given Latitude-Longitude or ZipCode or Postal
> Address.

> Q7: Identify models that are suitable for obtaining
> from concentration levels of certain pollutants at
> a specified time in a specific location.

This set of queries is most challenging because the models (even through semantic enhancement) do not directly take location/time and determine the pollutant concentration level at that time. With location co-ordinates the following information has to be inferred: (i) what are the different landforms available in the location (ii) what are the pollutant sources available near by the given coordinates and (iii) Are there more than one type of pollutant source in the given location (for example traffic and industrial source belonging to different category of the) and whether the models support multiple sources as input. With time the associated inferences are as follows: (i) Was air turbulence information available during that time, (ii) was vehicle induced turbulence information available during the specific time, (iii) what was the wind direction during the specified time, (iv) what are the weather conditions during the specified time at the given location. Once the individual inferences are performed the output has to be orchestrated in the query as either preconditions or IO to the AM function.

In order to evaluate this new functionality we synthesis a set of landforms and climatic conditions (for different time periods) considering a set of location co-ordinates. We randomly choose certain locations to contain different pollutant source types like volume, point, area and line. We allow the users to submit queries to this system by selecting a location co-ordinate. The users are not assumed to be the same people who created the semantic descriptions of the Analytic Functions. Our system automatically determines the nature of pollutant sources available in location, the climatic condition applicable for the time and allows aggregation of the input data to execute the appropriate AM to provide the concentration levels. When more than one AM is identified the choice is provided to the user and after appropriate selection of a AM, the AM is executed. Using this system we identified all relevant models (precision of 1 was achieved) for specific location/time were identified appropriately. Further it not only provides accurate model selection but also identifies and aggregates the necessary input required to execute the AM. This new feature is feasible to support due to semantic inferencing while with simple IO and functional description a precision/recall would be zero. It would require manual inspection with higher level of air quality AM expertise to determine the suitability of AM for queries ($Q6$ and $Q7$).

5   Related Work

5.1   Related Work

The literature on semantic web services and application integration, including science (also called e-science), is extensive, consisting of promising results and many challenges. To put it in perspective of this paper, we organize this section into two parts. The first relates to to application integration and the second to e-science.

Semantic web services [17] deals with both modeling of applications as services and their composition using semantic annotations. In the matching of annotations, [19] formalizes matching of web services from a directory based on various inexactness measures. In [16], the authors have identified the information that a Semantic Web Service must expose in order to fulfill the objective of automated discovery, composition, invocation and interoperation. While functional attributes have received attention, the non-functional attributes have not been much recognized in semantic web. They relate to performance, reliability and other user-acceptance issues. Since applications (software) are built to requirements, [8] describes how such requirements can be qualitatively arranged as goal structures and used to design systems. Their framework allows treating requirements as potentially conflicting or synergistic goals to achieve during the software development process. For service composition, many surveys capture the state-of-art [10,22,21] and open issues.

In science domain, services have been effectively demonstrated in domains like geology [5,6] and bioinformatics [12,1]. However, in prior work, the applications were already available as services while in our case, they may be simply described but need to be implemented or implementation be provided in some language not made into a services. Further, the constraints to determine application relevance are quite complex in environment but the specific data samples are small. Our initial results are promising and we are continuing to explore how previous techniques could apply in this new domain as we develop new ones.

6   Conclusion

Identification of appropriate analytical functions to determine air quality is one of the toughest problems in the domain of Environmental Monitoring and Management. According to USA's Environmental Pollution Agency (EPA), the leading organization in the domain of Air Pollution and Quality, it requires deep skills and domain expertise to identify the relevant model. In an attempt to simplify the discovery of analytical functions, we have proposed to utilize the semantic information describing model purposes, model outputs and model inputs. We showed that modeling Air Quality concepts and orchestrating them into requirements of the AM will not only simplify the identification of relevant models but also provides new search options through inferencing.

References

1. Adak, S., Srivastava, B.: Genome bioinformatics: Advancing biotechnology through information technology - part i: Molecular biology databases. Indian Journal of Biotechnology 1, 101–116 (2002)
2. Agarwal, V., Chafle, G., Mittal, S., Srivastava, B.: Understanding approaches for web service composition and execution. In: ACM Compute (2008)
3. National Environmental Agency: Singapore, nea., http://www.nea.gov.sg/
4. Akkiraju, R., Farell, J., Miller, J.A., Nagarajan, M., Sheth, A., Verma, K.: Web service semantics – WSDL-S. In: W3C Workshop on Frameworks for Semantics in Web Services (2005)
5. Altintas, I., Berkley, C., Jaeger, E., Jones, M., Ludascher, B., Mock, S.: Kepler: an extensible system for design and execution of scientific workflows. In: International Conf. on Scientific and Statistical Database Management (2004)
6. Altintas, I., Jaeger, E., Lin, K., Ludascher, B., Memon, A.: A web service composition and deployment frameworkfor scientific workflows. In: International Conf. on Web Services (2004)
7. Beychok, M.R.: Fundamentals of Stack Gas Dispersion: Guide. The Author (1994)

8. Chung, L., Nixon, B.A.: Dealing with Non-Functional Requirements: Three Experimental Studies of a Process-Oriented Approach. In: International Conference on Software Engineering, pp. 25–37 (1995)
9. APTI Virtual Classroom. So409: Lesson 6 plume dispersion and air quality modeling, http://yosemite.epa.gov/oaqps/eogtrain.nsf/HomeForm?OpenForm
10. Dustdar, S., Schreiner, W.: A survey on web services composition. International Journal of Web and Grid Services 1(1), 1–30 (2005)
11. EPA. Us environmental protection agency(epa), www.epa.gov/
12. Goble, C.A., et al.: Myexperiment: A repository and social network for the sharing of bioinformatics workflows. Nucleic Acids Research 38, Web Server issue, W677–W682 (2010)
13. Holmes, N.S., Morawska, L.: A review of dispersion modelling and its application to the dispersion of particles: An overview of different dispersion models available. Atmospheric Environment 40(30), 5902–5928 (2006)
14. IBM. Spss software, http://www-01.ibm.com/software/analytics/spss/
15. Irwin, J.S., United States, Earth Tech (Firm), Interagency Workgroup on Air Quality Modeling (U.S.): A user's guide for the CALPUFF dispersion model [microform] / U.S. Environmental Protection Agency, Emissions, Monitoring and Analysis Division ... [et al.]. The Division, Research Triangle Park, N.C (1995)
16. Lara, R., Lausen, H., Arroyo, S., de Bruijn, J., Fensel, D.: Semantic Web Services: Description Requirements and Current Technologies. In: International Workshop on Electronic Commerce, Agents, and Semantic Web Services (September 2003)
17. McIlraith, S., Son, T.C., Zeng, H.: Semantic web services. Intelligent Systems (IEEE) 16(2), 46–53 (2001)
18. World Health Organization. Air quality and health, www.who.int/mediacentre/factsheets/fs313/en/index.htm
19. Paolucci, M., Kawmura, T., Payne, T., Sycara, K.: Semantic Matching of Web Services Capabilities. In: First Int. Semantic Web Conf. (2002)
20. Pasquill, F.: The Estimation of the Dispersion of Wind borne Material. Meteorol. Mag. 90, 33–49 (1961)
21. Srivastava, B., Koehler, J.: Web Service Composition - Current Solutions and Open Problems. In: ICAPS 2003 Workshop on Planning for Web Services (2003)
22. Su, X., Rao, J.: A Survey of Automated Web Service Composition Methods. In: Proceedings of First International Workshop on Semantic Web Services and Web Process Composition (July 2004)
23. UK Department for Environment, Food and Rural Affairs (DEFRA), www.defra.gov.uk/

Relaxed Neighbor Based Graph Transformations for Effective Preprocessing: A Function Prediction Case Study

D. Satheesh Kumar, P. Krishna Reddy, and Nita Parekh

International Institute of Information Technology, Hyderabad, India
satheesh.kumar@research.iiit.ac.in, {pkreddy,nita}@iiit.ac.in

Abstract. Protein-protein interaction (PPI) networks are valuable biological source of data which contain rich information useful for protein function prediction. The PPI networks face data quality challenges like noise in the form of false positive edges and incompleteness in the form of missing biologically valued edges. These issues can be handled by enhancing data quality through graph transformations for improved protein function prediction. We proposed an improved method to extract similar proteins based on the notion of relaxed neighborhood. The proposed method can be applied to carry out graph transformation of PPI network datasets to improve the performance of protein function prediction task by adding biologically important protein interactions, removing dissimilar interactions and increasing reliability score of the interactions. By preprocessing PPI network datasets with the proposed methodology, the experiments conducted on both un-weighted and weighted PPI network datasets show that, the proposed methodology enhances the data quality and improves prediction accuracy over other approaches. The results indicate that the proposed approach could utilize under-utilized knowledge, such as distant relationships embedded in the PPI graph.

Keywords: Relax neighbors, pruning, graph transformation, data quality, protein function prediction.

1   Introduction

Proteins are involved in all important biological functions like construction of organs, proper production of enzymes, and maintenance of complex biological process in all living organisms. Role of a protein and the biological function in which it is involved is very important knowledge crucial for extracting knowledge about identifying disease specific proteins, functional modules, individual protein's functions [1,2], discovering new drugs, new crops with high productivity, pest resistance power, and synthetic bio-materials like bio-fuels etc [3].

Proteins of (PPI data) of many organisms are available due to advances in high throughput genome level screening experiments [4] like mass spectrometry. Many standard protein interaction databases such as GRID [5], DIP [6] are

S. Srinivasa and S. Mehta (Eds.): BDA 2014, LNCS 8883, pp. 114–128, 2014.

also available and freely accessible. Apart from many of these advantages PPI network data have some potential issues such as high level of noise in the form of false positive edges or spurious edges [11,12], incompleteness in the data due to absence of biologically valued edges. This is due to many previous experimental studies have considered specific individual decease specific proteins [11,12,14]. Such incompleteness prevents the state of the art protein function prediction algorithms to utilize entire network. There is a strong need for computational techniques to remove the noise and incompleteness.

Protein interaction datasets can be represented as an adjacency matrix of an undirected graph in which proteins are vertices and interactions are edges with their reliability scores. This graph representation can be exploited by applying graph analysis techniques to use underutilized knowledge to improve data quality. Efforts are being made by applying computational techniques based on neighborhood [7,18,31], global optimization [8], classification [1] and association analysis [9] on PPI network data for handling data quality challenges to improve the process of extracting useful knowledge like function or role of a protein and identifying protein complexes in a biological process. Finding reliability score of an interaction between two proteins by calculating similarity based on structural properties of protein interaction graph, mainly number of common neighbors, was the main theme used in many of the above approaches.

Existing approaches employ a specified similarity criteria to extract neighbors of a protein by fixing the value of corresponding similarity threshold. The similarity between two proteins is decided based on number of nodes in the common neighborhood. To improve the quality of reliability scores and avoid false positives, the similarity threshold to extract neighborhood is set to high value. It can be observed that, due to high similarity threshold, we may miss some genuine protein interactions between distantly or indirectly connected proteins. To exploit under-utilized knowledge to improve the performance, in this paper, we propose an improved method to extract similar proteins based on the notion of *relaxed neighborhood*. In this approach we relatively decrease or relax the value of similarity threshold to extract more neighbors with the expectation that increased number of proteins in the neighborhood establish more relationships among proteins. However, while forming a relationship between two proteins based on the number of common neighbors, there is a danger of getting more noisy relationships due to relaxed similarity. To avoid noisy relationships among proteins, we propose a pruning methodology which examines the properties of sub-graph formed by common neighbors of any two proteins. The similarity relationship is established, if the sub-graph satisfies the specified property in addition to specified threshold of number of common neighbors. By preprocessing of PPI network datasets with the proposed methodology, experiment results on both un-weighted and weighted PPI network datasets show that, the proposed methodology enhances the data quality and improves prediction accuracy over other approaches. The results indicate that the proposed approach could utilize underutilized knowledge, such as distant relationships embedded in the PPI graph.

In the next section we discuss related work. Section 3 we present the proposed approach. In section 4 we explain the experimental results. The last section contains summary and conclusions.

2   Related Work

Due to the availability of abundance of biological data lot of research work has been done on protein function prediction in the recent past. There are many approaches proposed in the literature. These approaches can be broadly classified into local approaches, global approaches, clustering based approaches, and association analysis based approaches [1].

Approaches based on neighborhood information of a protein like degree, density of common neighbors are proposed in the literature [16,9,17,31]. An approach based on indirect neighbors and topological weights was proposed which uses indirectly connected protein labels such as proteins at a shortest path 1 or 2 from each protein for function prediction [18]. These approaches use protein label information for protein function prediction. If the graph is incomplete or there are not many labeled neighbors present in the neighborhood, then it will be difficult to perform graph transformation for noise removal with these approaches.

Clustering [19] is used to identify dense regions of the PPI network data. Proteins interacting in the densest region are considered more functionally related. Functional annotations are assigned to annotate proteins based on neighborhood counting and majority voting methods [1, 4]. Protein function prediction is done by using protein functional labels present in the cluster [1]. Shared nearest neighbor clustering approaches use notion of common neighbors but use strict similarity criteria. All the above approaches use features like common neighbors between pair of proteins, graph cuts or some distance measures. In these approaches, each protein is forced to be a part of the cluster. Community extraction approaches are used to extract functional modules from PPI network data. These approaches do not consider relaxed neighborhood and common neighbor based subgraph properties based graph transformations for improving data quality by capturing similarity of distantly connected proteins.

Global approaches [8,9] are proposed where global features such as overall network connectivity information, frequent patterns can be used for protein function prediction. Some of the approaches are based on network flow. Nabevia et al [8] proposed functional flow algorithm where protein function prediction is done by using network flow based approach. In this approach unannotated proteins are annotated based on labels of directly connected proteins, distantly connected proteins with high edge weights. Panday et al [9] proposed association analysis based approach which uses support and Hconfidence measures to extract hyper clique patterns for noise removal on PPI data. There are classification approaches proposed for protein function prediction which also uses global relationships among proteins and protein functions [20]. These approaches do not fully utilize local structural information, notion of relaxed similarity to extract loosely related proteins, graph properties of common neighbor set subgraph.

Link structures are widely used in the co-citation analysis [26], bibliomet-
rics [27], social network analysis [28] for extracting useful link patterns to get
communities. Trawling approach [25] was proposed to extract potential cores of
communities from large collection web pages connected with hyper links. They
abstracted core of a community as a complete bipartite graph (CBG). An extension to it was to extract and relate hierarchy of communities with dense
Bipartite Graph (DBG) abstraction [22]. Current approaches in this area are
not exploited for the task of extracting and utilizing under utilized knowledge
and preprocessing protein interaction data to address data quality problems,
enhance functional content in the PPI network data to improve protein function
prediction.

The proposed approach in this paper is different as it extracts a set of loosely
related proteins based on the notion of relaxed neighborhood by relaxing similarity. It exploits the properties of subgraph of common neighbor protein set to
avoid potential formation of noisy edges.

3   Proposed Approach

In this section we explain motivation, basic idea and proposed approach.

3.1   Motivation

There is a scope to improve the performance of protein function prediction by
improving data quality trough including related proteins which are directly, indirectly connected in the neighborhood for evaluating reliability of protein interactions. In the Web environment, the notion of community is employed to identify
the related web pages for a given web page. The notion of community enables
the inclusion of related nodes with the appropriate relaxed criteria. Studies show
that there are group phenomena among proteins [1,18]. A protein interacts with
a group of other proteins at different times and at different places. Proteins
form protein complexes to carry out complex biological functions in pathways.
In addition to pairwise protein level knowledge, distantly connected protein set
properties are also very important. We can extract protein level and group level
knowledge by exploiting graph representation of protein interaction networks,
relaxing similarity to include set of distantly connected proteins in the relaxed
neighborhood and apply the graph properties of subgraph formed by a set of
common neighbor proteins of two proteins for pruning to establish a relationship between pair of proteins.

3.2   Basic Idea

There are several similarity criteria proposed in the literature. We explain the
basic idea of the proposed approach which can be used by considering any similarity criteria. In complex graph like a graph of PPI network, the neighborhood

of each node is extracted based on specified similarity criteria and similarly between two nodes is established based on number of common neighbors. Several methods in the literature vary based on type of similarity criteria.

We propose that there is an opportunity to improve the performance by extracting neighbors of a given node using relaxed neighborhood method, and deciding similarity between two proteins by examining the property of subgraph formed by common neighbors. To improve the quality of reliability scores and avoid false positives, we propose a relaxed neighborhood approach in which the similarity threshold to extract neighborhood is set to relatively low value. We relatively decrease or relax the value of similarity threshold to extract more neighbors with the expectation that increased number of proteins in the neighborhood establish more relationships among proteins. However, while forming a relationship between two proteins based on the number of common neighbors, there is a possibility of getting more noisy relationships due to relaxed similarity. To avoid noisy relationships among proteins, the property of sub-graph formed by common neighbors of any two proteins is examined. The similarity relationship is established, if the sub-graph satisfies the specified property in addition to specified threshold of number of common neighbors.

We explain the concept through an example. In *Figure 1*, if we use the number of edges on the path which connects two proteins as a distance measure to gather neighbors. We explain by setting the number of common neighbors as *4*, Jaccard similarity as *0.5* and link density as *0.5*. We explain by considering the cases of similarity between nodes pairs {*1* , *7*}, *and* {*2*, *6*}. With the path distance as 1, the neighbors of selected nodes are indicated as *1*={*1, 2, 4*}, *2*={*1, 2, 3, 7*}, *6*={*4, 5, 6*}, *7*={*2,4,7*}. With path distance as *2*, the neighbors of selected nodes are indicated as *1*={*1, 2, 3, 4, 6, 7*}, *2*={*1, 2, 3, 4, 5, 7*}, *6*={*1, 3, 4, 5, 6, 7*}, *7*={*1, 2, 3, 4, 6, 7*}. It can be observed that the number of nodes in the neighborhood has increased when path distance = *2* as compared to the path distance=*1*. If we check similarity between nodes *2* and *6* with distance *1* neighbors, we will get *0* because there are no common neighbors. If we consider distance=*2*, the common neighbor set is {*1, 3, 4, 5, 7*}, which is equal to *5*, Jaccard similarity is is *0.50* and common neighbor set subgraph link density is *0.5*. So we add an edge between nodes *2* and *6* which is not formed with path distance=*1*. For nodes *1* and *7*, with a path distance=*1*, Jaccard similarity is *0.5* there is a potential to add an edge, but with path distance=*2*, the number of common neighbors set is {*1, 2, 3, 4, 6, 7*}, Jaccard similarity is *1* and link density is *0.4*. We consider this edge unreliable as both values are below threshold. So, in this case, the edge is not added.

3.3   Approach

We observe that there is an opportunity to improve the performance by relaxing the criteria to extract neighborhood and at the same time to develop a mechanism to filter out non-potential similar pairs. So, the proposed approach consists of two concepts. One is an approach to extract neighborhood of a given protein

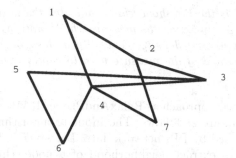

Fig. 1. A graph consists of 7 proteins. Each edge indicates an interaction between the corresponding proteins.

by relaxing the similarity criteria and another is a mechanism to filter out non potential similar pairs. We present the both concepts in the following.

- **Relaxed Neighborhood Extraction:** For each protein in PPI, the proposed relaxed neighborhood similarity *RSim* extracts loosely related proteins by relaxing the similarity criteria. As a result, the number of nodes in the neighborhood will be increased. Note that, the similarity criteria could be any of the popular similarity methods such as distance-based methods, link-based methods and so on. the definition for Rsim is given below.

   Definition 1. *RSim(u, p) : Let u and p be any two nodes in PPI graph. RSim(u,p)=true if* sim(u, p) $\geq$ simT. *Here,* sim(u,p) *denotes a similarity function and* simT *is the corresponding similarity threshold value.*

   Note that, in the preceding definition, the *sim* function could be any similarity function based on distance, path length, analyzing links and so on. The main objective is to give opportunity to more number of nodes to be part of neighborhood by relaxing *simT* value in a controlled manner.
- **Pruning Methodology:** For any two proteins u and p after extracting neighborhood using RSim, the issue is to establish relationship between u and p. Since, the neighbors extracted by relaxing similarity threshold, relatively there is increased number of neighbors as compared to non-relaxed case. Normally, existing approaches establish similarity based on the number of nodes common to u and p. In the proposed approach, we analyze the properties of neighbors common to u and p to establish relationship. If neighborhood set common to u and p do not satisfy the specified property, the relationship is not established. We capture this aspect using *Relate* measure which is defined as follows.

   Definition 2. *Relate(u, p) : Let u and p be any two nodes in PPI graph. The sets n(u) and n(p) indicate the neighborhood sets extracted using the function* Rsim. *Relate(u, p)=true if (* cn(u,p) $\geq$ cT *)* $\cap$ *(m(cn(u,p)) $\geq$ mT); Here, the notation* $\cap$ *indicates the logical AND,* cn(u,p) *denotes the common neighbour graph formed by the nodes and edges of neighbors common to u and p.*

The notation cT *is the threshold which indicates the number of nodes. The notation* m(cn(u,p)) *indicates the measure which captures the characteristics of* cn(u,p), *such as link density, centrality and so on. The notation* mT *indicates the threshold of the measure used to capture the characteristics of the graph.*

We call the proposed approach as RelaxAndPrune (RAP) aproach. The main steps of the algorithm are as follows. The input to algorithm is an undirected graph, like graph formed by PPI network data. Let $simT$, cT and mT indicate similarity threshold to extract neighborhood of a node, threshold of number of common neighbors, and threshold about property of sub graph formed by common neighbor set between two proteins.

Let there be n nodes in the graph, mT is the link density of the graph. Here the formula for mT is given as follows

$$mT = \frac{2m}{n(n-1)} \tag{1}$$

Here m indicates number of edges and n indicates number of nodes as a criteria. However, the proposed approach can be used by considering any other criteria.

1. For each node i, include j as neighbor if Rsim(i,j) $\geq$ **simT**.
2. For each pair of {i,j}, extract common neighbor set subgraph, link density of common neighbor set subgraph. Add an edge between i and j, if Relate(i, j) = true, that is if $((cn(i,j) \geq cT)$ AND link desity $\geq$ mT). Otherwise, remove an edge.

4   Experiment Results

In this section we explain the details of datasets, approaches evaluated, evaluation methodology, and experimental results.

4.1   Details of Datasets

We selected a popular and well accepted [1] **FunCat** functional annotation scheme for annotating proteins at a depth of 2 from MIPS database [23]. The dataset is related to protein interaction networks of S.Cerevisiae or budding yeast. Proteins are selected based on our selected functional annotations. The redundant interactions were removed like BA when AB is present and self interactions like AA. The dataset well known and well accepted.

We conduct experiments on two kinds of datasets. Table 1 provides the dataset name, size, number of proteins, the number of interactions. The DIP Core data set is created from a database of interacting proteins by selecting highly reliable 2315 proteins and 5413 interactions [30]. The other is Krogan et al's dataset. It is high-throughput and reliable dataset reported by Krogan et al.

Table 1. Details of datasets used for experiments

Dataset	No of Annotated Proteins	Number of Interactions
DIPCore	2315	5413
Krogan et al	2291	6180

[32] having 2291 proteins and 6180 interactions. In this dataset, the likelihood score of experimental reproducibility of each interaction was estimated by using various machine learning approaches. We used these likelihood scores as an edge weight.

4.2 Approaches Evaluated

We have conducted the experiments by comparing proposed approach with Hconfidence and common neighbor based approaches. The details regarding experimental set-up are as follows.

Common Neighbor Based Approach. The common neighbor based approach [21] assumes that proteins having more common neighbors will have function in common. Function prediction with common neighbors has obtained more accuracy than other approaches such as neighborhood counting and majority voting of the function label of directly connected proteins. Common neighbor based approach gives a stable level of accuracy as a number of common neighbors increases. In this approach edge is placed and high weight is given if there are more number of common neighbors else edge is not placed.

Hconfidence Based Approach. HConfidence measure [9] is used to extract similarity between two proteins based on the ratio of the common neighbors and minimum number of individual neighbors. In [9], it has been shown that Hconfidence approach performs better than common neighbour based approaches.

RelaxAndPrune (RAP). RAP is the proposed approach. For extracting relaxed neighbors, we used the path length between two proteins as similarity measure. After extracting neighbors, we have computed similarity between two proteins as follows. We first ensure that the similarity between the neighbor sets of the given proteins with Jaccard similarity measure is greater than the threshold value. Next, we also ensure the link density measure (Equation 1) of common neighbor graph is greater than the threshold value.

4.3 Evaluation Methodology

In this section we presented evaluation methodology, parameters used, details of function flow algorithm used to generate parameters. For experiments, we have taken Dipcore and Krogan PPI network datasets. For Dipcore dataset, we have created a unweighted and undirected graph and represented it as an adjacency matrix. For Krogan PPI network dataset, we have created a weighted and undirected graph and also represented as a adjacency matrix.

In the experiments, we have employed function flow algorithm [8] to predict the scores for protein label pairs. By taking these datasets as an input, we have preprocessed the graphs by adding edges, removing edges, and modifying edge weights using common neighbor, Hconfidence and the proposed RAP approach.

For each dataset, the transformed graph has been created. We gave original graph and then preprocessed graphs as an input to function flow algorithm and evaluated the performance of protein label score prediction through five fold cross validation based evaluation.

We explain the performance metric. In the context of protein function prediction, the definitions of precision, recall for multi label scenario are given in [9,21].

$$Precision = \sum_{i=1}^{n} \frac{K_i}{m_i} \tag{2}$$

$$Recall = \sum_{i=1}^{n} \frac{K_i}{n_i} \tag{3}$$

Here, k indicates the total number of proteins with known functional labels. For each protein i the notation m_i, indicated number of labels predicted by the algorithm, the notation n_i indicates the actual number of labels possessed by the protein. Precision denotes correct predictions out of all predictions performed, recall is correctly predicted annotations out of all annotations.

Accuracy of Top K Predictions: Biologists need small number of most promising functional predictions of proteins because of limitation of performing a small number of expensive experiments. In this scenario top k predictions will be more suitable than precision, recall or ROC curve. So, top k evaluation method was followed. We used global scoring methodology in which the top k protein label pairs with the highest scores are calculated. The prediction accuracy, precision is evaluated with respect to the known protein label annotations.

A graph with prediction accuracy is given on Y axes, the number of protein label pairs predicted on x axes is produced for different values of k. We used k=500 for both datasets.

$$\frac{Recall}{Precision} = \frac{k}{Tn} \tag{4}$$

The notation k is threshold for top k protein label pairs, Tn indicated total number of protein label pairs. The overview of function flow algorithm [8] as follows.

Function Flow Algorithm for Protein Function Prediction: Function flow algorithm [8] is based on network flow, which is a well known concept in graph theory. Many protein function prediction algorithms are proposed in the literature [9]. Among all function flow algorithm is based on well founded graph theoretic concepts and provides better accuracy. Due to this we have chosen this algorithm to test our hypothesis. Function flow approach is an iterative algorithm proposed for undirected graphs. With this approach annotated proteins are treated as sources or reservoirs and un-annotated proteins are treated as sinks. Network flow is simulated and controlled by edge weights treated as capacities which should not be exceeded. On each iteration a reservoir value or accumulated flow value is calculated for annotated proteins. Flow is allowed only from proteins with high reservoir value to proteins with low reservoir value. This flow is simulated for a fixed number of iterations for each function label. For each un annotated protein all the reservoir values are sorted in descending order and top 'K' values are obtained. Functional labels associated with those top 'K' values are assigned to un annotated protein. From the above discussion, we can say that more reliable edges with high reliability scores and accurate weights gives better protein function prediction. We restricted our evaluation to proteins labeled with at least one label at depth 2 in the FunCat hierarchy. Table 1 presents details. Different protein interaction datasets are preprocessed and then they are provided as an input to the function flow algorithm for evaluating their relative performance. We selected weighted, unweighted versions of datasets to show efficacy of the proposed method. We used the function flow algorithm in a fivefold cross validation method for predicting the score for each protein named with each label. This score can be converted into functional annotation. This process is explained next. We sorted all the scores of protein label pairs in descending order and select a threshold **K**. Scores above the threshold are predicted as annotations. Next metrics used for evaluating the performance of the algorithm is presented.

4.4   Experiment Results

We have conducted experiments by carrying out preprocessing with common neighbor, Hconfidence, and the proposed RAP approach. Table 2 summarizes pruning thresholds used, best value used for preprocessing. We used common neighbor based approach as a baseline method to compare with proposed RAP approach. We tried common neighbor thresholds 1, 2, 3 and selected 2 as a threshold which is best performing. Based on the number of neighbors and the degree of each node, Hconfidence is calculated. We selected Hconf=0, Min support as 2 for un-weighted graph, Hconf=0.20 for weighted graph. Using the proposed RAP approach, after examining the results with path length as 1, 2, 3 values, and we select the path length value as 2. We examined the results with Jaccard threshold value as 0.1. For pruning, we used link density threshold 0.021.

– **Experimental Results on DipCore Dataset:** Since DIP interactions are unweighted and we considered weight as 1 if there is an interaction, 0 if there is no interaction. The prediction scores returned by function flow algorithm through five fold cross validation based evaluation for all protein

Table 2. Details of experiments, the notation Hconf-bin is used for unweighted graph, Hconf-cnt is used for weighted graph

Method	Values tried	Best value
RAP	cT=0, 0.1, mT=0.020, 0.021, 0.022 simT=1, 2, 3	cT=0.1, mT=0.021 simT=2
Hconf	Hconf-bin=0, 0.1, 0.2,	0, 0.2
		Hconf-cnt=0, 0.1, 0.2
Common neighbors	Min common neighbors = 1, 2, 3	Min common neighbors = 2

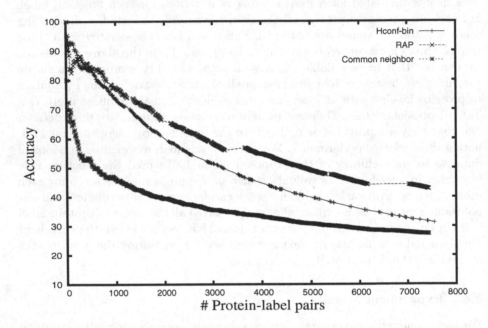

Fig. 2. Performance of all protein label pairs of DIPcore dataset

label pairs is given in Figure 2 and for top 500 protein label pairs is in Figure 3. It can be observed that the proposed RAP method improves the accuracy significantly over other approaches, because the proposed approach allows more nodes to participate in similarity computation by exploiting complex relationships. It can be observed that common neighbor method performs consistently low because of poor noise removal. Due to functional content improvement in the graph more reliable edges are added, better pruning of noisy interactions happened. Even on a reliable graph such as DIP core, RAP method gives good performance over Hconf.

- **Experimental Results on Krogan Dataset:** In this experiment, as Krogan dataset is weighted dataset, we have considered weights while computing Jaccard similarity. The prediction scores returned by function flow algorithm through five fold cross validation based evaluation for all protein label pairs

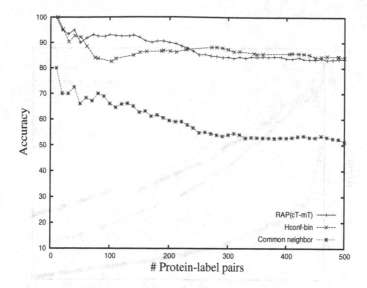

Fig. 3. Performance of top 500 protein label pairs of DIPcore dataset

is given in Figure 4 and for top 500 protein label pairs is in Figure 5. If we consider top 500 predictions, it can be observed that the proposed RAP approach performs better over the common neighbor and Hconfidence approaches, because the proposed approach utilizes more information in computing similarity.

Overall, the experiment results show that the proposed approach is able to improve the performance through effective preprocessing of PPI network data and extracts top k protein label pairs with high accuracy over other approaches.

5   Conclusions and Future Work

Protein function prediction by analyzing PPI network datasets is a research issue. In this paper we have proposed an improved approach based on the notion of relaxed neighborhood. After extracting the neighborhood of each protein, the similarly is computed by analyzing the common neighbor set subgraph. In this approach, a similarity of two proteins is determined by the corresponding neighborhood proteins which are extracted by relaxing the similarity criteria. It establishes the similarity between two proteins, if a group of sparsely connected proteins are common neighbors of the corresponding proteins. Experimental results show that graph preprocessing by relaxing neighbors produces more reliable and complete graphs when compared with other methods. The results on the popular Dipcore and Krogan's PPI network datasets show that proposed approach is able to improve the performance of protein function prediction over other approaches.

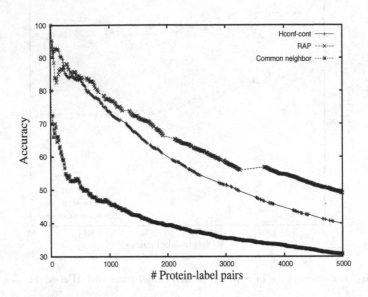

Fig. 4. Performance of all protein pairs of Krogan's dataset

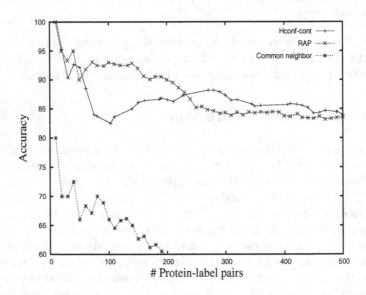

Fig. 5. Performance of top 500 protein pairs of Krogan's dataset

As a part of future work, we will conduct the detailed experiments. We will conduct experiments by considering alternative neighborhood extraction approaches by exploiting domain specific features. We will extend the proposed framework to improve the performance of graph-based knowledge/pattern extraction approaches.

Acknowledgements. We thank Dr Vipin Kumar, Dr Gaurav Panday for providing datasets, providing implementation of function-flow algorithm.

References

1. Pandey, G., Kumar, V., Steinbach, M.: Computational approaches for protein function prediction: A survey. Technical Report, Department of Computer Science and Engineering,University of Minnesota. 06-028 (2006)
2. Sharan, R., Ulitsky, I., Shamir, R.: Network-based prediction of protein function. Mol. Syst. Biol. 3, 88 (2007)
3. Chuang, H.Y., Lee, E., Liu, Y.T., Lee, D., Ideker, T.: Network-based classification of breast cancer metastasis. Mol. Syst. Biol. 3, 140 (2007)
4. Legrain, P., Wojcik, J., Gauthier, J.-M.: Protein protein interaction maps: a lead towards cellular functions. Trends in Genetics 17(6), 352 (2001)
5. Breitkreutz, B.J., Stark, C., Tyers, M.: The GRID: the General Repository for Interaction Datasets. Genome Biology 4(3), R23 (2003)
6. Xenarios, I., Salwinski, L., Duan, X.J., Higney, P., Kim, S.M., Eisenberg, D.: DIP, the Database of Interacting Proteins: a research tool for studying cellular networks of protein interactions. Nucleic Acids Research 30(1), 303–305 (2002)
7. Schwikowski, B., Uetz, P., Fields, S.: A network of protein-protein interactions in yeast. Nature Biotechnology 18, 1257–1261 (2000)
8. Nabieva, E., Jim, K., Agarwal, A., Chazelle, B., Singh, M.: Whole-proteome prediction of protein function via graph-theoretic analysis of interaction maps. Bioinformatics 21 (2005)
9. Pandey, G., Steinbach, M., Gupta, R., Garg, T., Kumar, V.: Association analysis-based transformations for protein interaction networks: a function prediction case study. In: KDD 2007: Proceedings of the 13th ACM SIGKDD International Conference on Knowledge Discovery and Data Mining, vol. 549, pp. 540–549 (2007)
10. Yona, G., Dirks, W., Rahman, S., Lin, D.M.: Effective similarity measures for expression profiles. Bioinformatics 22(13), 1616–1622 (2006)
11. von Mering, C., Krause, R., Snel, B., Cornell, M., Oliver, S.G., et al.: Comparative assessment of large scale datasets of protein protein interactions. Nature 417, 399–403 (2002)
12. Hart, G.T., Ramani, A.K., Marcotte, E.: How complete are current yeast and human protein interaction networks. Genome Biology 7, 120 (2006)
13. Deng, M., Sun, F., Chen, T.: Assessment of the reliability of protein protein interactions and protein function prediction. In: Pac. Symp. Biocomputing, pp. 140–151 (2003)
14. de Silva, E., Thorne, T., Ingram, P., Agrafioti, I., Swire, J., et al.: The effects of incomplete protein interaction data on structural and evolutionary inferences. BMC Biology 4, 39 (2006)

15. Girvan, M., Newman, M.E.J.: Community structure in social and biological networks. Proc. Natl. Acad. Sci. USA 99, 7821–7826 (2002)
16. Brun, C., Chevenet, F., Martin, D., Wojcik, J., Guenoche, A., et al.: Functional classification of proteins for the prediction of cellular function from a protein-protein interaction network. Genome Biology 5, R6 (2003)
17. Pandey, G., Manocha, S., Atluri, G., Kumar, V.: Enhancing the functional content of protein interaction networks. CoRR abs/1210.6912 (2012)
18. Chua, H.N., Sung, W.K., Wong, L.: Exploiting indirect neighbours and topological weight to predict protein function from protein-protein interactions. Bioinformatics 22, 1623–1630 (2006)
19. Pereira-Leal, J.B., Enright, A.J., Ouzounis, C.A.: Detection of functional modules from protein interaction networks. Proteins 54(1), 49–57 (2003)
20. Yu, G., Domeniconi, C., Rangwala, H., Zhang, G., Yu, Z.: Transductive multi-label ensemble classification for protein function prediction, pp. 1077–1085. KDD (2012)
21. Lin, C., Jiang, D., Zhang, A.: Prediction of protein function using common-neighbors in protein-protein interaction networks. In: Proc. IEEE Symposium on BionInformatics and BioEngineering (BIBE), pp. 251–260 (2006)
22. Reddy, P.K., Kitsuregawa, M.: An Approach to Relate the Web Communities through Bipartite Graphs. In: WISE 2001, pp. 301–310 (2001)
23. Ruepp, A., et al.: The FunCat, a functional annotation scheme for systematic classification of proteins from whole genomes. Nucleic Acids Research 32(18), 5539–5545 (2004)
24. West, D.B.: Introduction to Graph Theory. Prentice Hall (2001)
25. Kumar, R., Raghavan, P., Rajagopalan, S., Tomkins, A.: Trawling the Web for emerging Cyber communities. In: 8th WWW Conference (May 1999)
26. Garfield, E.: Cocitation analysis as a tool in journal evaluation. Science, 178 (1772)
27. Broder, A., Kumar, R., Maghoul, F., Raghavan, P., Rajagopalan, S., Stata, R., Tomkins, A., Wiener, J.: Graph structure in the Web: experiments and models. In: 9th International WWW Conference (May 2000)
28. Scott, J.: Social Network analysis: a handbook. SAGE Publications (1991)
29. National Center for Biotechnology Information, http://www.ncbi.nlm.nih.gov
30. Deane, C.M., Salwinski, L., Xenarios, I., Eisenberg, D.: Protein interactions: two methods for assessment of the reliability of high throughput observations. Mol. Cell. Proteomics (2002)
31. Camoglu, O., Can, T., Singh, A.K.: Integrating multi-attribute similarity networks for robust representation of the protein space. Bioinformatics Journal (July 2006)
32. Krogan, N.J., et al.: Global landscape of protein complexes in the yeast Saccharomyces cerevisiae. Nature (2006)

Remote Interactive Visualization of Parallel Implementation of Structural Feature Extraction of Three-dimensional Lidar Point Cloud

Beena Kumari, Avijit Ashe, and Jaya Sreevalsan-Nair

Graphics-Visualization-Computing-Lab,
International Institute of Information Technology Bangalore,
26/C Electronics City, Bangalore 560100, India
beena.kumari@iiitb.org, {avijitashe,jnair}@iiitb.ac.in
http://www.iiitb.ac.in/GVCL/index.html

Abstract. Lidar (Light detection and ranging) is a popularly used technology to collect dense and precise data of topographic structures of the surface of the earth. In this paper, we have proposed a user assisted remote visualization system for extraction and visualization of structural features obtained from point cloud data obtained from lidar data. The sharp feature lines such as crest lines, edges of buildings, ravines, ridges are known as structural features of point cloud. Our work includes: (a) parallel implementation a topology-based algorithm for extraction of structural features from lidar point cloud using GPGPU (General Purpose Graphics Processing Unit) computing, and (b) using a LAN (Local Area Network)-based server-client architecture to achieve remote visualization.

Keywords: 3D interactive visualization, lidar point clouds, feature extraction, stochastic methods, GPGPU (parallel) computing.

1 Introduction

The lidar (Light Detection and Ranging) technology is heavily used for flood-modeling and similar applications, such as, bathymetry, geomorphology, glacier modeling, etc. Point cloud or positional vector data can be processed from raw lidar data. We have focused on using topology-based algorithm to extract features from such point cloud data. Point cloud datasets are generally visualized directly as a point cloud or in a derived form as a mesh. Our primary focus in this work is to reduce the point cloud to essential points, which encode structural features.

Additionally, raw lidar datasets contain noise due to instrumental errors and atmosphere. Hence, proper pre-processing and filtering are necessary for both managing large scale dataset as well as for denoising. We have used the heuristic algorithm proposed by Keller et al. [3] to denoise and to extract the structural features from lidar datasets. However, the algorithm is highly computationally intensive and the serial implementation of the preprocessing module is time

S. Srinivasa and S. Mehta (Eds.): BDA 2014, LNCS 8883, pp. 129–132, 2014.

consuming. In order to speed up the preprocessing for data simplification, we have exploited the data-parallel characteristic of the algorithm and implemented the parallel algorithm using GPGPU computing techniques [5] using CUDA libraries [4].

Visualization of lidar data has been motivated by target users of this visualization who are domain experts from national laboratories or similar organizations. The target users additionally expressed the interest to access such a visualization across a LAN (local area network), so that the data resides in a server and the interactive displays may be enabled on thin clients. The server-client architecture has been implemented using a product called ThinLinc [2].

2   System Description

The processing system consists of three modules: (a) a server, which performs all the back-end operations, (b) a thin-client, which performs all the front-end operations and presents the user interface, and (c) the virtualization interface, which is the transporting mechanism that enables the communication between the front-end and the back-end.

2.1   Back-End (Server) Computing and Rendering

We have implemented the server-client architecture using HP Z400 Workstations running Ubuntu 12.04.1 LTS as the server. We have used OpenGL libraries for the graphics support, and CUDA libraries for GPGPU computing. As an alternative to our OpenGL rendering application, we have used Paraview, which is an open-source cross-platform data analysis and visualization application developed by Kitware. Our OpenGL-based rendering tool is dedicated towards point cloud visualization while Paraview has extensive capabilities to perform visual data analytic operations like slicing, masking the point cloud, etc. Using Paraview, we have performed meshing using the greedy surface triangulation algorithm on a point cloud dataset with normals, where the outcome is a triangle mesh based on projections of the local neighborhoods. While currently we have focused on point cloud visualization, we are interested in extending the modes of visualization to other standard techniques, such as, surface and volume visualizations.

2.2   Virtualization Interface

User interactive sessions and real-time image manipulation and rendering calls for a very efficient VNC (Virtual Network Controller). TigerVNC is the most suitable protocol for this endeavor, especially for our application that includes integration of high-end graphics framework. ThinLinc is a Remote Linux Desktop Server developed by Cendio AB [2], entirely built on open-source. It provides users with centralized remote access to both Windows and/or Linux desktop applications simultaneously and is a secure, cost-effective and freely install-able solution for up to ten concurrent users. Overall, ThinLinc meets with our requirements for this work. TigerVNC is integrated into ThinLinc.

OpenGL Rendering Support. VirtualGL is one of the most robust solutions to allow OpenGL applications to be ported via a VNC. Traditionally the 3D data is transmitted over the network to be finally rendered on the client machine. VirtualGL is designed to direct the OpenGL commands and the 3D data to the graphics accelerator on the application server, as in our case, and direct the rendered images across to the client machine. Thus it enables users to interact in real-time using even hand-held devices as the front-end device is used only for display and user interactivity. VirtualGL is optionally integrated with Thin-Linc, and we have additionally tested our system with VirtualGL integration on ThinLinc.

2.3 Front-End (Thin-Client) UI

The thin-client essentially has X libraries installed to enable the display and interaction for the GUI (graphical user interface). We have tested the system on the following thin-client configurations: (a) Ubuntu 12.04.1 LTS running on Dell XPS as well as on Dell Inspiron 3520, and (b) Windows 7 Ultimate running on ASUS X42.

3 Implementation

We have implemented the algorithm proposed by Keller et al. [3] to detect and extract structural features in the objects in the lidar point cloud. These features include corners, border-, crease-, or ridge-lines in the form of a feature graph. The algorithm consists of the following modules: (a) outlier removal/denoising, (b) stochastic point classification, (c) smoothening of feature values, and (d) feature graph construction. The serial implementation of the algorithm is slow, owing to its computationally intensive nature. Hence we have implemented the algorithm in parallel using CUDA [4] and PCL [6] libraries, and tested on an Intel Xeon(R) processor at 3.2GHz quad-core, NVIDIA GeForce GTX480. Table 1 compares the performances of the serial and parallel implementation.

4 Experiments and Discussions

We have analyzed the results of the serial and parallel implementations of the algorithm proposed by Keller et al.[3] for various datasets. Table 2 gives the classification of the points into different classes - curved, planar, critical curved and critical planar and also shows the extent of reduction in, each case, upon computing the feature graphs. We can see that the coplanarity of the points in the data set, and the reduction of the point data set are positively correlated

5 Conclusions and Future Work

We are reducing the point cloud data procured using lidar technology using derived information of inherent structural features, using a heuristic topology-based algorithm [3]. We have observed that several modules in the algorithm

Table 1. Comparison between serial and parallel implementation of structural feature extraction algorithm [3]

Datasets [1]	test2	test1	galvestone	mscstsc	spring2
# Points	11765	33703	99660	160101	201474
Serial (in CPU seconds)	2.02	6.73	48.76	84	144.63
Parallel (in CPU seconds)	0.07	0.25	1.36	2.25	4.76
Speedup	28.85	26.92	35.85	37.33	30.38
Datasets [1]	srsota	N144835	N440375	autzen-stadium	spring1
# Points	386530	431276	497536	693895	926276
Serial (in CPU seconds)	262.96	369.95	439.51	460.21	2529
Parallel (in CPU seconds)	4.34	9.4	10.04	18.91	38.73
Speedup	60.58	39.35	43.77	24.33	65.29

Table 2. Point Cloud Classifications and percentage of reductions

Dataset	Point Classification					Point Reduction	
	# Points			# Critical Points		# Points	%age
	Total	Curved	Planar	Curved	Planar	(Reduced)	Reduction
Leica	1,642,660	106,654	1,589,446	321	2,744	108,276	93.40
Stadium	693,895	54,174	633,524	1,358	1,376	139,435	79.90
Dragon	437,645	35,975	420,319	215	220	146,882	66.43
Stanford Bunny	35,947	14,621	29,893	29	29	20,337	43.43

are embarrassingly parallel and hence, we have used CUDA for efficient parallel implementation. To enable the server-client architecture for this application, we have used ThinLinc to deploy the application over a LAN. Such a setup enables organizations with a dedicated LAN to use our tool to explore and analyze the lidar datasets without any hardware upgrade. We will be further working on the tracking of the features across time-series data-sets to find temporal patterns.

References

1. Butler, H., Loskot, M.: Sample Datasets at libLAS - LAS 1.0/1.1/1.2 ASPRS LiDAR data translation toolset (2013), http://www.liblas.org/samples/
2. Cendio AB: Thinlinc (2014), https://www.cendio.com/products/thinlinc/
3. Keller, P., Kreylos, O., Vanco, M., Hering-Bertram, M., Cowgill, E.S., Kellogg, L.H., Hamann, B., Hagen, H.: Extracting and Visualizing Structural Features in Environmental Point Cloud LiDaR Data Sets, pp. 179–193. Springer, Heidelberg (2010)
4. NVIDIA: CUDA (2013), https://developer.nvidia.com/category/zone/cuda-zone
5. Owens, J.D., Houston, M., Luebke, D., Green, S., Stone, J.E., Phillips, J.C.: GPU computing. Proceedings of the IEEE 96(5), 879–899 (2008)
6. Rusu, R.B., Cousins, S.: 3D is here: Point Cloud Library (PCL), May 9-13 (2011)

Elections Again, Twitter May Help!!!
A Large Scale Study for Predicting Election
Results Using Twitter

Pulkit Mehndiratta[1], Shelly Sachdeva[1], Pankaj Sachdeva[2], and Yatin Sehgal[3]

[1] Jaypee Institute of Information Technology
{pulkit.mehndiratta,shelly.sachdeva}@jiit.ac.in
[2] Maharaja Agrasen Institute of Technology
pankajsachdeva14@gmail.com
[3] Northern India Engineering College
ytn.sehgal@gmail.com

Abstract. With the widespread acceptability of social media and networking sites, more and more people are coming forward to express their
views and opinion about multidisciplinary topics. Nevertheless, people do
talk about politics and politicians on these sites enabling us to explore
the opportunity of monitoring and mining the opinions of large number
of politically active population in real time. In this paper, we have tried
to analyze a very famous micro-blogging online social network Twitter,
where users read and write millions of short messages known as tweets
on a variety of topics every day. We conducted the content analysis of
over 0.25 million tweets containing a reference to either a political party
or a politician for election which were being conducted in April 2014 in
India. Our analysis for tweets indicated a very close connection to the
parties and political position of politicians thus, can conceivably imitate
the offline landscape of the elections. Finally, we discuss the use of micro-
blogging message content as a legitimate pointer of political sentiments
and develop suggestions for the future research.

Keywords: Online Social Networks, Twitter, Elections, Opinion mining, Sentiment Analysis, Text analysis.

1 Introduction

Online social networks (OSNs) sites have given all new dimensions to the world
of communication among each other. People now share continuous streams of
messages to share their interests, indications, discuss activities, status, opinions,
latest trends etc. Thus, these networks can be used as an effective means to
testify trends and popularity in topics ranging from entertainment, politics, environment, economic and social issues and many more. Out of all these trending
topics, politics is the one which has received much of the attention.

The total expenditures on politics are increasing steadily not only in India
but in other countries too. In few countries there are no restrictions on the

S. Srinivasa and S. Mehta (Eds.): BDA 2014, LNCS 8883, pp. 133–144, 2014.

corporate funding in political campaigns. Thus, much prominence is on this new cheap platform for making connections with the voters as well as for promotion purposes. It is also the other way around; people are more and more involved with the political processes online.

1.1   The Twitter and World!!!

Twitter [2] is an online social network and micro-blogging site, which enable users to send and read the short 140-character text messages known as tweets as shown in Fig.1. The registered users can post and read tweets, while the unregistered users can only read the post. If a person likes the post of one user then he/she can share that particular tweet from own profile and the process is known as retweet. Twitter users have developed metadata annotation schemes which, shows that it compress substantial amount of information into a comparatively tiny space.

Fig. 1. An example of a Tweet

Currently, the user base of twitter is over 650 million [1] worldwide, with almost 0.13 million new users joining daily. Over a billion, new tweets are being posted every month on this site from a wide range of interest groups. Twitter is in the list of the most visited sites according to Alexa ranking [3], and has been described as *"the SMS of the Internet"*. Its large scale and streaming nature makes Twitter an ideal platform for monitoring events in real time.

From the political view, Twitter can be considered as an opening for up-to-date status updates, allowing campaigns, candidates and people to respond to particular news and political events. It can be the media which is used to conduct marketing and spread the popularity of any candidate like wildfire.

1.2   Sentiment Analysis

Sentiment analysis [24]is an application of text analysis techniques for the identification of subjective opinions in text data. It normally involves the classification of text into categories such as "positive", "negative" and in some cases "neutral". Over the past few years, there is an increasing demand for sentiment analysis tools by companies willing to monitor customers or employees opinions of the company and on its products and services. To fulfil the increasing demands for such kinds of tools, more and more researchers and companies are releasing products to perform sentiment analysis, many of them claiming to be able to perform sentiment analysis of any type of document in every domain.

Unfortunately, experience has shown us that, an "out-of-the-box" sentiment analysis tools working across domains does not exist yet. The main reason sentiment analysis is so difficult is that words often take different meanings and are associated with distinct emotions depending on the domain in which they are being used. There are even situations where different forms of a single word will be associated with different sentiments. For example, in customer feedback the word "improved" was associated with positive comments, but "improve" is more often used in negative ones. All sentiment analysis tools rely at varying degrees on lists of words and phrases with positive and negative connotations or are empirically related to positive or negative comments.

1.3   The Indian Lok Sabha

The Lok Sabha or House of the People is the lower house of the parliament of India. It comprises of the people elected from 543 constituencies, chosen by direct elections on the basis of adult suffrage. Indian Lok Sabha is the largest running democracy in the world and unlike other countries the candidates are from more than two parties which is the usual practice.

Analyzing and gathering information from the tweets to find out the patterns and predicting results for an election thus, makes it very interesting and off-the-shelf.

Following the same trend the authors have conducted a study to extract and analyze the information from Twitter. The aim of the study is twofold. First, we examine whether Twitter can be used as a media for online political reflection by seeing at how people use it. Secondly, can it reflect the actual offline political sentiment and orientation in a rational manner? Also, the authors try to answer the question if it is possible to automatically classify tweets; this is because newsfeeds or tweets tends to have an informational character, while user-generated messages usually have a communicational purpose.

2   Background and Related Work

Over the past few years, the exponential growth of Twitter has received attention from people from all the communities i.e., researchers, business analysts,
and stock brokers. As Twitter provides constant and continuous stream of real-
time updates from around the globe, much research is in detecting noteworthy,
unexpected events. Some of the studies in the past show that how Twitter can
help in detection of influenza outbreak [4], seismic events like earthquakes [5] and
the identification breaking new stories [6][7][8]. These applications are similar to
what is done in streaming data mining efforts paying attention on various other
media outlets.

Not only in field of monitoring the real-time activities, Twitter can be used as
an powerful spam detection tool [9][10][11]. Another very prominent and emerging field of research is area related to sentiment analysis on the huge Twitter
corpus. Luca et al. presented a well structured and detailed study of how to
sense any trending topic on Twitter [12]. In another study, Goorha and Ungar
used Twitter data to develop sentiment analysis tool for a famous company Dow
Jones to detect and predict significant emerging trends relating to specific products and companies [13]. While our findings gave us an insight on why and how
people use micro-blogging services, but they have not explored the use of this
new communication service in context to specific instances like: public relations,
political debate or political orientation of the people.

Given the scenario, the question arises on what grounds these predictions
have been made. Could it be simply a mere coincidence or is there a reason
why general trends are as accurate as specific demographics? Another important
question which arises is, should these methods and techniques can be applied for
future predictions also hence generalized enough to perform globally.

Thus, the authors have conducted the study in a different mode altogether.
The current study is trying to analyze a huge corpus of 0.25 million tweets
which may contain some gibberish data and trying to find out which party is
more popular in terms of positive and negative popularity. As results of general
elections were still awaited at the time of writing this paper, our study has
given us some important heads-up about the three major parties participating
in General Lok Sabha Elections 2014 in India. We have been able to find out
which party is most popular and has better chances of winning as compared to
others in the elections.

The rest of the paper is organized as follows: section 3 discusses about the
methodology and data set used for the analysis purpose. In section 4, we have
shown various results and finally, section 5, provide the conclusions of our study
conducted on the data.

3   Methodology

The most important goal of the paper is to find whether Twitter can actually
predict the winner of any election, irrespective of demographic details and dependency. Taking the interaction based perspective on political communication,

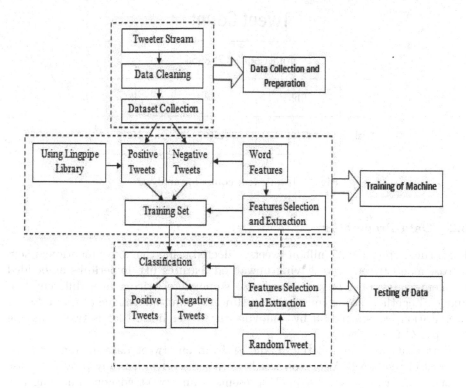

Fig. 2. The System Architecture

the current research developed the algorithm to support above hypothesis. Figure. 2 shows the methodology used, which is divided into various phases. These phases are further divided into sub-steps, description of which is as follows.

3.1 Data Collection

We had streamed tweets from twitter using streaming API (Application Programming Interface) named as twitter4j [25]. The keywords for collection of tweets were selected after a careful study of tweets during a week's interval. The most frequently used words were: bjp, modi, cong, gandhi, aap, and kejriwal. All the tweets are related to these keywords, related tweet were selected and stored in our database for the analysis purpose. We collected over 0.25 million tweets for this analysis and evaluation purpose as (shown in Fig.3), out of which BJP had around 0.1 million tweets, AAP had around).12 million tweets and remaining number belonged to INC.

Tweet Count		
Total Count = 258964		
	BJP	AAP	INC
Total	97900	125000	55000

Fig. 3. Total tweet count for analysis

3.2 Data Preparation

From the corpus of 0.25 million tweets collected using the above mentioned keywords, many tweets were having unwanted features like hyperlinks embedded
in them, many were retweets, and also some were written in a different language altogether. For example, like in Hindi, the mother tongue of India. So for
preparation, we started off by removing the hyper links, that is tweets having
"http://" tag in them.

Post that, we removed the # and @ from our tweet dataset and retained
the rest of the tweet. Also, stop words were removed which only provided noise
in the system as they are equally present in any tweet without inferring any
sentiment. Also, regular expression (regex) was used to remove other special
character present in the tweet.

After thorough cleaning, the authors were only left with the important words
that would constitute our feature vector for training the machine.

Example: If the initial tweet was like as given below: BJP's PM candidate
Narendra Modi addresses BJP supporters at a rally in Lakhimpur, Uttar Pradesh
#India2014 http://t.co/WDrn5bQbeK

The modified tweet after the cleaning of the http:// and stop words is: BJP
PM candidate Narendra Modi addresses BJP supporters rally Lakhimpur Uttar
Pradesh #India2014

Similarly, @ and # from the data stream had been removed.

3.3 Training of Machine

The most important part of sentiment analysis classification is to provide a good
quality training data set. For this, we manually handpicked about 10 positive
and 10 negative tweets for all 3 political parties: (BJP, Congress and AAP).

After creating a classifier using these handpicked tweets, we first created a
set of 1000 tweets for each party to form the test data and classified them as
positive or negative using the LingPipe library [16]. Lingpipe is a software library for natural language processing (NLP) implemented in Java. Lingpipe's

application programming interface (API) is tailored to abstract over low-level implementation details to enable components such as tokenizers, feature extractors, or classifiers to be swapped in a plug and play fashion. Lingpipe contains a mixture of heuristic rule-based components and statistical components, often implementing the same interfaces, such as chunking or tokenization. More specific aspects of Java coding relating to text processing, such streaming I/O, character decoding, string representation, and regular expression processing is implemented in this library.

The technique we used from this lingPipe library is based on logistic regression algorithm [17] and use conditional probability to classify tweets. After detecting the sentiment for 1000 tweets for each of the 3 political parties, we further added these tweets to the training data thus, increasing the size of our training data set. This recursive process was repeated by selecting unique tweets from our dataset of 0.25 million until all of them were classified either as positive or negative. Fig. 3, provides the steps involved during training and testing the database.

A tweet could be positive for one party and the same tweet could be negative for another. So, our system uses different classification cluster for all 3 parties and it is possible that one tweet could be in positive training data set for one party and negative in another one.

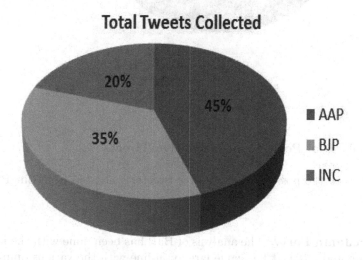

Fig. 4. Total Tweets Collected

3.4   Testing of Data

Once the machine has been trained to classify the positive and negative tweets we have passed the remaining tweets to the trained machine to automatically classify them as negative and positive tweets.

4   Analysis and Results

After collecting tweets over a period of time, using our search query terms such as, (bjp, modi, cong, gandhi, aap, kejriwal), we have found out that most discussed party is AAP on twitter.com with more than 0.12 million tweets out of 0.25 million we collected. BJP hold the second slot next to AAP, with little less than 0.1 million tweets out of 0.25 million we collected. Congress was least discussed or trended over twitter with about 50,000 tweets as shown in Fig.4. The count of tweets is approximating, not exact.

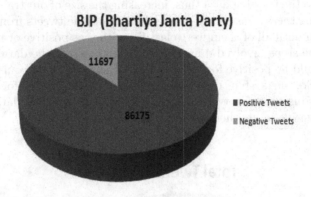

Fig. 5. Results for BJP

4.1   Analysis of Positive and Negative Tweets

The clusters of the positive and negative tweets, generated from LingPipe is as follows:

Bhartiya Janta Party. The analysis of BJP has been done with the keywords used as bjp and modi. It has came largely in-line with the various opinion polls conducted and predicted during the period when the data collection was taking place as shown in Fig.5.

Aam Aadmi Party. From a total of 0.12 million tweets, the analysis shows (Fig.6) that about 40% are in favor of AAP, and 60% tweets are in negative. This is may be due to resignation of their party leader after being elected as the chief minister of country capital New Delhi in Assembly elections December 2013 [18].

AAP (Aam Aadmi Party)

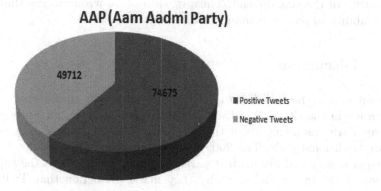

Fig. 6. Results for AAP

INC (Indian National Congress)

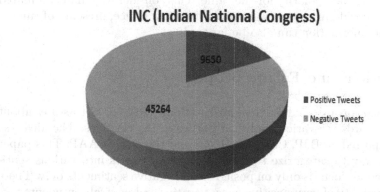

Fig. 7. Results for INC

INC (Indian National Congress). From a total of 50,000 tweets, the analysis shows that about 18% are in favor of Indian National Congress, and a huge amount of about 82% tweets have been classified as negative or against the party (Fig.7.). The overall results of the analysis somehow show the real offline picture of the actual trend. Thus, it will not be wrong to say that Twitter has an ability to sense the trends in real-time.

5   Discussion

Twitter may have ability to predict elections with an amazing amount of accuracy in the elections in the past. O'Connor et al. in 2010 presented an early paper [19] discussing about the feasibility of using Twitter data as a substitute for traditional polls. The 2009 German elections were predicted in [14]. This paper clearly had two distinct parts, firstly it discuss about the superficial analysis of the tweets and secondly they made prediction that Twitter data was rather close to that of actual polls. Similar kind of study was conducted for 2010 US Congressional elections discussed in[15] and also in [20, 22] for the Singapore General Elections and Irish General Election respectively. On the contrary, studies [21, 23] are there in the past which actually question the presumed predictive powers of Twitter data regarding elections.

All the studies conducted in the past were *post-hoc*. They used the term prediction could have been made, and unnecessary to say the chances of getting negative result is minimum when the actual election results are already out. But, here in this study the results were actually predicted before the actual results were declared. There is no standard and commonly accepted way to count the votes on Twitter, thus we had tried to concentrate on the hash-tags only to collect the tweets and evaluate polarity for the same. This consideration had eliminated the chances of considering every tweet as trustworthy. Hence, presence of rumors and misleading information can also be ignored.

6   Conclusions and Future Scope

This report focus on 14th Indian General Elections, 2014 so, as to study about the support towards the various political parties in the elections. The idea was to get tweets posted on BJP, Congress (INC) and the newbie AAP. This paper summarizes a way to categorize twitter messages or tweets into various sentiments, however our focus is only on positive and negative sentiments only. There could be a possibility of very positive, very negative and neutral sentiments too.

AAP is said to be a party born because of the social media and the present form is greatly influenced by the social media. Also, it is widely reported that political parties have established special cells in their workforce to manage their social media image. So, we collected large number of tweets for better analysis and the results showing us a bigger picture of which political party is getting maximum support online.

The analysis shows that, AAP trended the most on twitter followed by BJP and Congress. However, the analysis of positive and negative tweet among them show a different picture, i.e., the political party who was most trending on Twitter doesn't imply that it had more supporters. In our analysis, BJP got 82% positive tweets whereas 18% negative tweets. AAP ratio stood 60% in negative and 40% in positive and Congress was very badly perceived with 82% in negative.

As these analysis were done before declaration of the actual election results (2014) declared. We have done a comparative study of our results with the actual results of 14th Indian General Elections, 2014 available in appendix.1 at the end of this paper. In India, the internet reach is only around 16% and out of those, only 35% people actually talk about politics on social media thus, this analysis doesn't simply imply that the resulting seats outcome are exactt imitation of what is popular on Tweeter. Our analysis show that BJP is the most favored party on Twitter i.e., maximum positive support is there for them twitter.

In future, we try to perform the deeper analysis of the same corpus generated, as a lot of work can still be done for the neutral and highly biased tweets. At the same time, we cannot neglect the usage of ironic and sarcastic words or sentences which actually has a different meaning as compared to what is written.

A   Appendix

As discussed earlier, the paper was written when result was still awaited, but when we compared the out analysis results with the actual election results, they were in very close correspondence to each other. Our results showed that BJP is the most popular party in positive sense and this party actually won the elections with complete majority winning 282 seats out of 543.

AAP (Aam Aadmi Party) which got some steam in the past few months was only popular on twitter but not at all when it comes to real world scenario. Thus, saying that this party has been promoted and got publicity by social media is not wrong completely.

Lastly talking about the oldest of all, the Indian National Congress Party (INC) was the one which was most popular in negative sense and same was replicated when the results arrived. Not only they lost the elections, rather they lost majority to form a government by just winning 44 seats out of 543. Thus, it is not wrong to say that Twitter can actually imitate the real world scenarios very closely.

References

1. Twitter statistics, http://www.statisticsbrain.com/twitter-statistics/
2. Twitter, https://www.twitter.com
3. Alexa: The Web Information Company, http://www.alexa.com
4. Culotta, A.: Towards detecting influenza epidemics by analyzing Twitter messages. In: Proc. of the 1st Workshop on Social Media Analytics, pp. 115–122. ACM (2010)
5. Sakaki, T., Okazaki, M., Matsuo, Y.: Earthquake shakes Twitter users: real-time event detection by social sensors. In: Proc. of the 19th International Conference on the World Wide Web, pp. 851–860. ACM (2010)

6. Phelan, O., McCarthy, K., Smyth, B.: Using Twitter to recommend real-time topical news. In: Proc. of the 3rd ACM Conference on Recommender Systems, pp. 385–388. ACM (2010)
7. Petrovic, S., Osborne, M., Lavrenko, V.: Streaming first story detection with application to Twitter. In: Proc. of NAACL, Citeseer (2010)
8. Sankaranarayanan, J., Samet, H., Teitler, B., Lieberman, M., Sperling, J.: Twitterstand: News in tweets. In: Proc. of the 17th ACM SIGSPATIAL International Conference on Advances in Geographic Information Systems, pp. 42–51. ACM (2009)
9. Yardi, S., Romero, D., Schoenebeck, G., Boyd, D.: Detecting Spam in a Twitter network. First Monday 15(1), 1–13 (2009)
10. Wang, A.: Don't follow me: Spam detection in twitter. In: Proceedings of the International Conference on Security and Cryptography, SECRYPT 2010 (2010)
11. Kwak, H., Lee, C., Park, H., Moon, S.: What is Twitter, a social network or a news media? In: Proc. of the 19th International Conference on World Wide Web, pp. 591–600 (2010)
12. Aiello, L.M., Petkos, G., Martin, C., Corney, D., Papadopoulos, S., Skraba, R., Gker, A., Kompatsiaris, I., Jaimes, A.: Sensing Trending Topics in Twitter. IEEE Transactions of Multimedia 15(6), 1268–1282 (2013)
13. Goorha, S., Ungar, L.: Discovery of significant emerging trends. In: Proc. of the 16th ACM SIGKDD International Conference on Knowledge Discovery and Data Mining, pp. 57–64 (2010)
14. Tumasjan, A., Sprenger, T., Sandner, P.G., Welpe, I.M.: Predicting elections with twitter: What 140 characters reveal about political sentiment. In: Proc. of 4th ICWSM, pp. 178–185. AAAI Press (2010)
15. Livne, A., Simmons, M., Adar, E., Adamic, L.: The party is over here: Structure and content in the 2010 election. In: Proc. of 5th ICWSM (2011)
16. Lingpipe Library, http://alias-i.com/lingpipe/
17. Peduzzi, P.: A simulation study of the number of events per variable in logistic regression analysis. Journal of clinical epidemiology 49(12), 1373–1379 (1996)
18. President's Rule imposed in Delhi after Arvind Kejriwal's resignation, http://www.ndtv.com/article/india/president-s-rule-imposed-in-delhi-after-arvind-kejriwal-s-resignation-484458
19. O'connor, B., Balasubramanyan, R., Routledge, B.R., Smith, N.A.: From Tweets to Polls: Linking Text Sentiment to Public Opinion Time Series. In: Proc. of the 4th International AAAI Conference on Weblogs and Social Media, pp. 122–129 (2010)
20. Skoric, M., Poor, N., Achananuparp, P., Lim, E.-P., Jiang, J.: Tweets and Votes: A Study of the 2011 Singapore General Election. In: Proc. of the 45th Hawaii International Conference on System Sciences, pp. 2583–2591 (2012)
21. Metaxas, P.T., Mustafaraj, E., GayoAvello, D.: How (Not) To Predict Elections. In: Proc. of PASSAT/SocialCom, pp. 165–171 (2011)
22. Bermingham, A., Smeaton, A.F.: On Using Twitter to Monitor Political Sentiment and Predict Election Results. In: Proc. of the Workshop on Sentiment Analysis where AI meets Psychology, SAAIP (2011)
23. GayoAvello, D.: Don't turn social media into another 'Literary Digest' Poll. In: Communications of the ACM (CACM), pp. 121–128 (2011)
24. Pang, B., Lillian, L.: Opinion mining and sentiment analysis. Foundations and trends in information retrieval 2(1-2), 1–135 (2008)
25. Twitter4j, http://twitter4j.org/en/

Detecting Frauds and Money Laundering: A Tutorial

Girish Keshav Palshikar

Tata Consultancy Service Limited,
54B Hadapsar Industrial Estate, Pune 411013, India
gk.palshikar@tcs.com

Abstract. The purpose of this tutorial is to provide an introduction to
the general area of frauds to analytics scientists and professionals and
discuss some analytics techniques used in their detection. We focus on
frauds in insurance, stock markets and on money laundering. There are
survey papers [1], [2], [3] and books [4], [5], [6], [7], [8] that discuss
various analytics techniques for fraud detection in general. However, they
do not survey analytics for stock market frauds and money laundering.
Another important contribution is that we also discuss some open areas
and research problems in the field.

1 Introduction

It is well-known that modern businesses lose hundreds of billions of dollars annually due to *frauds* - aka *economic offences, financial crimes, scams or white-
collar crimes*[1]. The problems of frauds are particularly severe in today's computerised, web-connected, mobile-accessible, and cloud-enabled business environments. Frauds are prevalent across all industry segments: banking and financial
services, insurance, telecom, retail, transportation, and government organizations. An FBI report[2] states that the insurance industry in the US, which consists of over 7000 companies and collects over \$1 trillion in premiums, loses about
\$40 billions annually in frauds in the non-health insurance sector alone. As per
The Coalition Against Insurance Fraud, the insurance industry in the US lost
about \$80 billion in 2006 to fraud. Estimates for frauds in the US for health-care
insurance vary from \$51 to \$115 billion annually[3]. Continued prevalence of such
mal-practices on a large scale can have disastrous long-term consequences not
only for the businesses involved but also for the customers, investors, financial
institutions, the government and economy, in general.

Oxford Advanced Learner's Dictionary defines a fraud as "an act of deceiving
illegally in order to make money or obtain goods". Indeed, in frauds, groups of
unscrupulous individuals manipulate or influence the activities of a target business, with the intention of making money or obtaining goods through illegal or

---

[1] PriceWaterhouseCoopers LLP *2009 Global Economic Crime Survey.*

[2] http://www.fbi.gov/stats-services/publications/insurance-fraud

[3] Sources quoted in http://en.wikipedia.org/wiki/Insurance_fraud

S. Srinivasa and S. Mehta (Eds.): BDA 2014, LNCS 8883, pp. 145–160, 2014.
© Springer International Publishing Switzerland 2014

unfair means. A fraud cheats the target organization of its legitimate income and results in a loss of goods, money and even good will and reputation. A fraud often employs illegal and always unethical or unfair means. A fraud often consists of many *instances* or *incidents* involving repeated transgressions using the same method. The instances of a same *type* of fraud are similar in contents and "appearance" but usually not identical. Some fraud instances are opportunistic, one-time occurrences committed by single individuals acting alone, while most are systematically committed and frequently repeated occurrences by organized groups. Frauds are different from closely related crimes such as corruption, theft, industrial espionage, sabotage, IP violations, and robberies. But even within a particular organization, the full scope of what exactly constitutes frauds is not always clear. In some frauds, there is internal complicity or compromises from some employees of the organization. In such cases, it is difficult to distinguish frauds from losses due to incompetence, procedural lapses, accidents, mismanagement, wrong decisions or business risks.

While modern businesses may be more prone to frauds, they also have highly automated business IT systems which collect vast amounts of data regarding almost all kinds of transactions and activities within a business. With the advent of data warehousing, it is now possible to access not only the current business data but also historical data. Clearly, evidence for frauds and fraudulent activities is partly hidden in these enormous quantities of data. It should be possible, using data analysis techniques, to discover knowledge to perform effective fraud detection and to prevent losses and to help bring the culprits to justice. It is of course necessary to combine the data-based evidence with evidence from other sources, such as personal interactions, witness statements, other investigations etc., to identify fraud perpetrators and build an effective legal case. Moreover, automated fraud detection can also reduce workload on fraud surveillance managers and investigators, allowing faster closure, through a focused selection of suspicious cases and better use of the investigators' time.

The need to apply analytics to detect frauds is well-understood and extensively researched. There are good survey papers [1], [2], [3] and even books [4], [5], [6], [7], [8] that discuss applications of various analytics techniques to fraud detection. The purpose of this tutorial is not to review but to provide an introduction to the general area of fraud detection to analytics scientists and professionals and discuss some of the analytics techniques used in their detection. We focus specifically on frauds in insurance, in stock markets and on money laundering. These books and surveys do not cover applying analytics to detect stock market frauds and money laundering. As another important contribution, we discuss some open areas and research problems in the field.

2 Fraud Typologies

The types of frauds differ across different domains, though there are common elements (misinformation, padding, collusion etc.). There is a need to develop a comprehensive taxonomy of frauds that goes across various domains. Some examples of types of frauds in automobile and P&C insurance are:

Types of Frauds in Automobile Insurance

Claim for damages/injuries that actually did not happen

Claim for treatments not actually received

Claim for injuries not related to automobile accidents

Lying about the passengers or driver present at the time of accident

Lying about facts of accident (location, timing, role of other parties)

Misreporting wage losses due to injuries

Excessive costs (padding) for damages/injuries than actually paid

Deliberately create an accident or create damages

Claim repairs for damages that are not covered under the policy

Deliberate collusion among service providers to inflate costs

File multiple claims for the same accident/injury

File claims when claimant was not involved in the accident/injuries

Types of Frauds in Property and Casualty Insurance

Inflated inventory of lost property (e.g., property never owned or sold previously)

Phony burglary

Staged (deliberate) fire

Claim for a sunk boat, but the boat never actually existed

3   Analytics for Automobile Insurance Fraud Detection

[9] empirically compared several binary classification algorithms (under different paremeter settings) for their efficacy in detecting fraudulent automobile insurance claims: logistic regression, decision trees, k-nearest neighbours, support vector machines, naive Bayes, tree augmented naive Bayes and Bayesian learning multilayer perceptron neural networks. The percentage correct classification and mean area under ROC curve were used as comparison criteria. The dataset consisted of 1399 personal injury protection (PIP) claims settled in Massachusetts in 1993. The PIP cover in Massachusetts covers 100% of medical expenses and 75% percent of wage losses, up to the limit of $8000. The claim amounts varied from a minimum of $10 to a maximum of $104487, with average $2761 and median $1765. There were two types of features: 25 red-flag features were binary fraud indicators computed using domain-knowledge based rules, and the other 12 features were more directly available data elements. Some red-flag features are: Single vehicle accident, Claimant in an old, low-value vehicle, Property damage inconsistent with accident, Non-emergency treatment was delayed, Was difficult to contact/uncooperative etc. Some non-flag features are: Age, No. of days from the start of the one-year policy to accident, No. of days from accident to reporting, No. of days from accident to first outpatient treatment, Ambulance charge, Whether claimant is represented by attorney. One important consideration in feature selection was that the values of the features should be available early enough in the claim processing life cycle. Each claim had two labels: one was a 10-point expert suspicion score (scale of 0 to 9, 0 indicating no fraud) and the

other was a 5-point expert verbal assessment giving an indication of the type of fraud (1:probably legitimate, 2:excessive treatment, 3,4: opportunistic fraud 5:planned fraud). None of the binary classification algorithms was found to be a clear winner, most had comparable accuracies; linear logistic regression and linear least-squares SVM were found to be slightly better.

Once the suspicious claims are identified using a trained classifier, there is a need to select those which will actually be *investigated*. Moreover, the insurance company may use specialized auditing strategies to pick additional claims for investigations (e.g., high-cost claims, claims recommended by expert claim reviewers). However, investigations are costly, time-consuming and need expert skills. Investigations help in confirming fraud, detecting frauds as well as a deterrent. Some methods of investigations are: independent medical examination, medical audit, activity check, site investigation, recording statements, getting sworn statements etc. In an operational setting, the effectiveness of any fraud detection mechanism can be measured by the % of investigated claims which were found to be truly fraudulent. Other measures of operational effectiveness could be % of costs recovered or % of convictions won for claims which went to the litigation stage and so on. Further, there is a need to compare various investigation strategies in terms of their costs and effectiveness. [10] used a subset of 1091 claims in the same dataset as used in [9], out of which 358 claims were investigated and about 75% them confirmed as fraudulent. However, not all investigation strategies were equally effective. They trained logistic regression models from the data and found that SIU referral was the most effective (23.8% refuted claims came from it), followed by activity checks (14.3%), medical audits (13.8%), site investigations (5.7%) and recorded statements (3%).

With the availability of a plethora of classification algorithms, it is feasible to use many of them and use ensemble methods, such as bagging and boosting, to combine their predictions into a final decision. Boosting builds a sequence of classifiers in a fixed number of iterations, each classsifer trained on the a random sample selected from the same training data. The weight of each sample (and hence higher probability of being selected) varies from iteration to iteration. Basically, samples that were incorrectly classified in the previous iteraiton receive higher weight in the next iteration. The predictions of these classifiers are combined using a weighted voting scheme. [11] built a sequence of 25 Naive Bayes classifiers using boosting and uses them to identify suspicious claims on the automobile insurance claim fraud data. [12] developed a fraud detection method to predict criminal patterns from skewed data, which uses a single meta-classifier (stacking) to choose the best base classifiers (naive Bayes, C4.5, and back-propagation), and combines their predictions (bagging) to improve cost savings.

Frauds are relatively rare and so in a a labeled dataset, known cases of frauds tend to be a minority class, where often with only a few % of cases are labeled as frauds. Two broad classes of techniques have been developed to handle learning from skewed (imbalanced) datasets: cost-sensitive classifiers and sampling. Since failure to detect fraud cases has a much higher cost, one can adjust the costs attached in predicting the class label. Sampling techniques such as SMOTE produce similar synthetic cases from known fraud cases, which is useful to reduce the class imbalance. There is a need to explore the quality of results produced by such techniques.

4   Frauds in the Stock Market

Stock markets are large virtual financial markets where people trade *stocks* (*shares*) of companies (securities) listed on the *stock exchange*. Modern stock exchanges are fully automated, where all aspects of the trading are computerised and records of all transactions maintained in the databases. The aggregate size of the 52 regulated stock exchanges across the world (total market capitalization) was $55 trillion as on Dec. 2012 (http://www.aei-ideas.org accessed 06-Sep-2014). Given the money involved, it is not surprising that the stock market is a target of frauds. Among many types of stock market frauds, we discuss here two main types of frauds: circular trading and insider trading.

4.1   Insider Trading

The trading price of a company's shares is affected by many factors, such as those related to the business of the company (e.g., new contracts) as well as decisions of the company (e.g., new products, alliances, financial results and dividend declarations). *Insiders* are people who have access to such confidential, proprietary or privileged information about the company, before it is known to the general public. Insiders can make large fortunes by using the privileged information to take decisions to buy or sell the company's shares. There are several laws that regulate *insider trading*; e.g., the Insider Trading Sanctions Act of 1984 and the Insider Trading and Securities Fraud Enforcement Act of 1988 in the US. It is closely watched by regulatory bodies. Insiders include company's owners, directors and senior management as well as large share-holders. *Temporary insiders* also include employees (e.g., enginers) and partners (e.g., accountants), if they have access to such information. Moreover, insiders often pass on such information to other people (e.g., friends, brokers) who then use it to make profits in trading. Such *tippees* are also treated as insiders. People (e.g., hackers) who steal and use such information are also insiders. www.sec.gov lists many insider trading cases, including some high-profile ones.

Detecting insider trading is challenging because one needs to identify insiders or their proxies, identify the nature of the privileged information and establish that there was "significant" trading by insiders in the duration from the information became available to them till just before the privileged information became public. Thus one needs to mine the trading data, company data as well as public-domain textual documents (news, press releases).

[13] analyzes raw data from three sources: option trading, stock trading, and news for a six-month time period to detect possible occurrences of insider trading in options. The 27 features from stock and option data measured five high-level symptoms about how a company's trading deviated from its own normal

behavior. 8 features were extracted from business news, where one challenge was how to separate important news from unimportant one. An data element was labeled positive if within the next two weeks after the date, there was significant news that led to a sizable increase in stock price. A jump in a stock's price, even taken together with high volumes and other trading irregularities, may look suspicious but does not always involve (or prove) insider trading. The authors compare decision tree, logistic regression, and neural net results to results from an expert rule-based scoring model.

4.2 Circular Trading

In order to ensure that the trading in stock markets is "fair", the authorities usually have laws and guidelines (e.g., Securities and Exchange Board (SEBI) guidelines in India) to be followed by all participants in stock market transactions. In price manipulation, a group of individuals acts together to artificially (and with a view to profit making) influence the price of a security. Typically, this is done to increase the price of the so-called "penny stocks". Price is influenced usually by circulating false information or by creating an artificial demand for the security. To achieve the latter, some people form an informal group, and circulate a number of shares among themselves in a large number of trades. They keep increasing the price in these trades, thereby forcing an increasing trend in the price as well as attracting other people. When the overall trading price rises sufficiently, the traders in the group "exit" by selling their original shares. Since the price rise was not tenable, the price crashes back to its original level or below. Such *circular trading* or *collusion* within a group is one particular type of unfair trading. See details of a circular trading case on the SEBI web-site[4], where 3 people traded among themselves (total trading volume > 200000 shares) using synchronized orders and increasing sequence of prices to help increase the share price from Rs. 340 to Rs. 450 within 6 trading sessions. In one part, 2 people within the group made 19 trades between themselves in 7 minutes, where the price difference between the first and the 19th trade was Rs. 24.6.

Several factors make collusion detection a challenging task: lack of training data, the vast size of the trading datasets, unknown identities and even size of the collusion set, unknown start and end times for collusion trading, complex strategies that may have been used in collusion trades, possibly multiple simultaneous collusions. Moreover, each individual transaction is perfectly legal and the collusion becomes apparent only when the transactions are grouped appropriately. It is challenging to establish that the collusion trades actually influenced the price and volume of trades and that the colluders made profits thereby and may need significant follow-up investigations such as interviews, analysis of delivery and payment records etc. In addition, various instances of the circular trading have similar but not necessarily identical behaviour (or traces) in the trading databases. Hence, there is no guarantee that ability to detect one oc-

---

[4] SEBI Order CO/109/ISD/12/2004 against DSQ Holdings dated 10th December 2004
http://www.sebi.gov.in

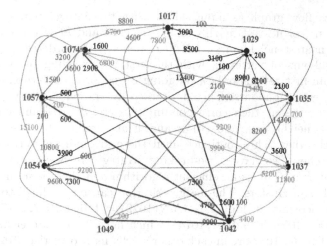

Fig. 1. An example stock flow graph

currence would be sufficient to detect the other occurrences. Note that we must necessarily examine the *sequences* of trades to discover collusion.

We describe one unsupervised approach [14] for collusion detection. First, the trading data for a particular company within a specified time-period is converted into a directed, edge-labeled graph, where the vertices are traders, there is an edge from u to v if there was at least one trade with u as seller and v as buyer, and the edge label is a feature vector that summarizes the trades form u to v. The feature vector can be as complex as desired and may include total no. of trades, total no. of shares traded, total value of the trades, timespan of the trades etc. Such a *stock flow graph* can have parallel edges in opposite directions but not self-loops nor isolated vertices, and it can be disconnected. Figure 1 shows an example stock flow graph of 9 vertices an 49 edges, where the edge label consists of only one feature: total no. of shares traded.

Next, they formalize the notion of collusion set as a connected subgraph H of the stock flow graph G, such that the traders within H trade much more heavily within themselves as compared to their trading with others outside H. The notion of heavy trading among themselves can be formalized in various ways. Note that there is no requirement that the subgraph H be a clique. Similarity of the notion of the collusion set as a subgraph with the notion of a community should not be lost, which should open up the possibilities of using community detection techniques on the stock flow graph.

Next, the authors develop several graph clustering algorithms to detect collusion sets (which is a connected subgraph) in the stock flow grpah, because graph clustering algorithms are based on the idea that a vertex is likely to belong to the same cluster as that of its *nearest* neighbours.

They adapted the well-known shared nearest neighbour graph clustering algorithm to the problem of candidate collusion set detection. This algorithm takes a stock flow graph as input, along with two integer parameters k and kt (k is the no. of nearest neighbours to be used and kt is the minimum no. of common nearest neighbours). $kNN(u)$ denotes the ordered sequence of k nearest neighbours of vertex u, with the first vertex being the nearest to u. The algorithm first retains only k outgoing edges for each vertex, one to each of its k nearest neighbours. Then it iteratively compares every pair of vertices u and v, merging their corresponding clusters, if: (i) u and v share more than $kt \leq k$ neighbours; and (ii) u and v are among the k-nearest neighbours of each other. This continues until there are pairs of vertices that can be merged. They also have a tie-braking strategy in vase u can be merged with either v or w. This is an efficient algorithm and there is no need to specify the number of clusters. The merge criterion can be modified to take other features into acocunt. In Figure 1, $4NN(1029) = \{1054, 1037, 1017, 1042\}$ and $4NN(1037) = \{1054, 1029, 1017, 1074\}$, which have 2 common elements (1054, 1017) and 1029 and 1037 are in each other's 4NN list; so vertices 1029 and 1037 can be merged. For the example stock flow graph, the following candidate collusion sets are detected under different settings:

$k = 5, kt = 2, \{1029, 1035, 1037, 1042, 1074\}$
$k = 6, kt = 4, \{1029, 1037, 1042, 1074\}$
$k = 7, kt = 5, \{1029, 1037, 1042, 1057, 1074\}$

Clearly, there is a strong case to consider the common vertices among these $\{1029, 1037, 1042, 1074\}$ as the final candidate collusion set, which can then be investigated using other methods. This group traded a total of 76000 shares among themselves, which is 28.6% of the total shares traded (266000). The group traded a total of 140100 shares with the rest of the people; thus their ratio of internal to external trading is $76000/140100 = 52.2\%$, which is a noticeably high fraction, compared to a random subset of 4 traders. The group's total trading volume is 216100, which is 81.2% of the total, indicating that the group consists of heavy traders in this security during this period.

The authors adapt another well-known graph clustering algorithm called mutual nearest neighbour and also propose a new algorithm specifically tailored for finding collusion set. These 3 algorithms can be run with different parameter settings, which yields a sequence of candidate collusion sets. The authors compute a numeric evidence value for each candidate collusion set generated, and then use the well-known Dempster-Shafer Theory of Evidence Combination to compute the final candidate collusion set(s), which have the highest evidence against them. [15] propose a two-step visualization-based solution for fraud detection in stock markets, where first they use 3D treemaps to monitor the real-time stock market performance and to identify a particular stock that produced an unusual trading pattern. Then they perform social network visualization to conduct behavior-driven visual analysis of the suspected pattern, identify the entities involved in the fraud, and further attack plans.

5 Money Laundering: An Overview

5.1 What Is Money Laundering?

Money laundering (ML)[5] is a serious problem for the economies and financial institutions around the world [17], [18], [19]. As a recent example, a global bank paid a fine of $1.9 billion to the US government in a large ML case[6]. Financial institutions get used by organized criminals and terrorists as vehicles of large-scale money laundering, which presents them with challenges such as complying with regulations, maintaining financial security, preserving goodwill and reputation, and avoiding operational risks like liquidity crunch and lawsuits. With its connections to organized crimes as well as terrorist financing, ML has become a serious issue worldwide and has been receiving considerable attention from national governments and international bodies such as the UN, IMF and the World Bank. Many countries have stringent laws against ML. See the Egmont Group (http://www.egmontgroup.org) which is an international consortium of the national Financial Investigation Units (FIUs) of over 100 countries, the Wolfsberg Group of global banks (www.wolfsberg-principles.com) and an international consortium of 33 countries called the Financial Action Task Force (FATF) (www.fatf-gafi.org) for international cooperation regarding ML.

ML refers to activities performed with the aim of enabling the use of illegally obtained (*dirty*) money for legal purposes, while hiding the true source of the money from government authorities. Dirty money often comes from predicate (underlying) crimes such as drug trafficking, illicit arms trades, smuggling, prostitution, gambling, corruption and bribery, fraud, piracy, robbery, currency counterfeiting etc. It may also come from incomes of legal businesses that need to be hidden for evading taxes. ML enables the conversion of cash from the underground (shadow) economy into monetary instruments of the legal economy. Hiding the true source of the dirty money (activities, locations, people, and organizations) is a crucial requirement for ML.

Internet, online banking and investments, and new financial services have made it easier to conduct international transactions as well as to hide or steal true identities. Since the types and volumes of legal transactions are very large, and often the individual transactions involved in an ML episode appear superficially legal, it is not easy to detect entire ML episodes, particularly because criminals may be using innovative and unknown schemes. Hence automated data mining techniques are needed, which can be coupled with experts' domain knowledge about financial transactions and criminal investigations for effective ML detection. Detecting, investigating and proving ML requires much domain-knowledge, in banking, business, legal etc. Key issues in investigating an ML episode are: identifying the modus operandi, identifying monetary instruments and institutions involved, identifying parties and beneficiaries involved, proving the unlawful origins of the money, tracing the transactions, proving liabilities and

---

[5] This section is based on [16].

[6] http://www.reuters.com/article/2012/12/11/us-hsbc-probe-idUSBRE8BA05M20121211

intent, proving violations of particular laws, and prosecution. Anti-ML (AML) regime also includes preventive actions (audits, employee training) and detection and regulatory compliance steps (suspicious transaction report, currency transaction report), etc. Several commercial products for AML are available. As a rough estimate, financial institutions spent about $500 million in 2012 on AML products and services.

5.2   Methods of Money Laundering

The various schemes used for ML have 3 broad stages. In the *placement* stage, the funds derived from criminal activities are introduced into financial systems, they are moved around for varius purposes and are converted to less suspicious monetary forms. In the *layering* stage, the funds are passed through many institutions and jurisdictions to disguise their origins (i.e., hide audit trail); this is done through mechanisms such as wire transfers, and newer monetary instruments and are usually international operations. In the *integration* stage, the funds are brought into an economy where they appear to be legitimate business earnings. ML schemes often require participation and cooperation of many partners (witting or otherwise), such as associates from illegal businesses, couriers, attorneys, accountants and banks and other financial institutions. Several building blocks are commonly in any specific ML scheme. We examine some of them, as they are more amenable for detection through analytics.

Nominees and Proxies: Nominees are real persons with real identity who are used to perform various tasks as part of the ML scheme: operate bank accounts, buy and sell assets (real estate, jewelry, cars). Nominees may also be designated to hold various posts (directors, officers) in the companies (shell companies, offshore corporations) involved in the ML scheme and carry out tasks related to these posts. Nominees are trusted agents who are themselves aware that they are being used as part of a ML scheme; e.g., relatives, friends, associates in an illegal activity, professionals (lawyers, accountants). Nominees are usually in close proximity to the real ML principals, so as to be able to receive instructions and execute tasks. A proxy is like a nominee but has a fake or stolen identity. The objective in using nominees/proxies is to conceal and protect the main persons in the ML scheme (e.g., by hiding the true ownership of assets). It is sometimes easy to identify a nominee or a proxy, particularly for ownership of an asset, because tax or utility bill payments may show that it is the real ML principal who actually uses the asset.

Front Businesses: A front business is a good vehicle for ML because, dirty money can be mixed with the genuine revenue of the business, assets can be held in the name of the business (thus shielding the ML principal) and the principal can easily control the business through nominees or in person. Moreover, the business premises may be shared for the illegal activities of the ML principal. To be suitable the front business should be such that it can be done with few people and it naturally handles large amounts of cash, so that dirty money can be better hidden inside the business cash flow. A service business is better than goods business, because there is a possibility of creating fake customers for the services. Businesses such as hotels, restaurants, bars, nightclubs, liquor stores, theaters, travel agencies, currency exchanges, money transfers, import-export agencies, shipping companies, and equipment rentals are widely used as front businesses. An offshore front business can provide an additional layer of anonymity, because some countries have strict laws that protect the privacy of business records. A front business is usually not a completely dummy business; it does conduct at least some valid business activities. Auditors can help in finding whether the business is genuine or whether it has many dummy business transactions that did not actually take place. Auditors can examine the traces of business process (e.g., business records, bank records, government records etc.) and internal controls within the process, to locate transactions that have falsified records or those which have not followed the business process properly.

Smurfing: Smurfing is a placement strategy to change cash into other forms, where the ML principals use people (*smurfers*) to conduct cash transactions (make deposits, buy cashier's cheques) below the reporting limits across many banks and accounts.

Shell (pyramid) Companies: These are legally formed and inter-linked dummy companies (some of which may be overseas) without any real business activities, usually staffed by nominees/proxies. They are used in layering as fronts to move funds around, buy assets and to conceal true ownership.

Identifying nominee/proxy bank accounts, bank accounts of shell companies, to detect smurfing operations and to identify front businesses all pose significant challenges to analytics.

5.3 Analytics for Money Laundering Detection

There are essentially two perspectives in ML detection, differing in terms of the data available. One is from the perspective of a central government surveillance organization, which has data (mainly CTR, STR etc.) from *all* financial institutions in the country. The other is from the perspective of a single organization (e.g., a bank), which has the data only of transactions within its boundaries and has also to comply with AML regulations.

Senator et al. [20] from FINCEN have created the FINCEN AI system (FAIS) that links and evaluates reports of large cash transactions to identify potential ML and has been in operation at FINCEN since 1993. The objective is to detect previously unknown, potentially high-value entities (transactions, subjects, accounts) for possible investigation. The model supports three belief levels: Reported, Accepted, and hypothesized. The reported transactions are at the belief level of Reported. These transactions are consolidated in clusters. Summary data like Subject and Account clusters, computed from the sets of reported transactions, represent the next belief level (Accepted). At these levels, certain derived attributes are computed that are necessary for evaluating the data-driven suspiciousness based on information discovered by analysts, including the linkages among the clusters. The highest level of belief (Hypothesized) is used for higher level abstractions like cases and patterns. FAIS has integrated the Alta Analytics NETMAP link analysis package, which uses the wagon-wheel displays. FINCEN uses both wagon-wheel displays and traditional link-and-edge charts for analysis. FAIS has attempted use of techniques like Case-based Reasoning (CBR) and data mining (nearest neighbor, decision trees), which were not very successful due to the lack of many labeled examples. Even unsupervised learning algorithms were found to be not so reliable because of difficulties in deriving appropriate features due to poor data quality and the need for background knowledge. These techniques were found to be useful as knowledge engineering aids. Analysts have used FAIS to generate the suspiciousness score and evaluated the subjects through research and analysis of the data available from all the sources for development of valid leads. These leads are then fully researched and analyzed by the law enforcement agencies. FINCEN uses the feedback from these agencies to make improvements in the system.

We now examine some analytics challenges for ML detection from perspective (2). Identifying bank accounts which are likely to be held by nominees/proxies is quite challenging. The main difficulty is that we have to identify nominee/proxy accounts only from the banking transactions data. Generally, other sources of side information, such as credit card data or tax documents are not available. Detailed analysis of cash coming in and other types of transactions (e.g., overseas money transfer) can provide good features to distinguish such accounts from other normal accounts. Looking at the instruments used for the outgoing money as well as the actual goals of these transactions (e.g., to buy assets such as real estate) can also help. Time-based patterns of transactions within the accounts and with the outside world are helpful. Another important challenge is to identify which businesses are being used as fronts for ML, from the banking (or other) transactions of all businesses. In general, movement of money is crucial in ML and analytics should focus on identifying suspicious money movements.

Given the fact that only a few entities will be known as having participated in ML, active learning can be used to reduce the need for labeled data. At each step, an active learning method selects one data point for manual labeling by the user and uses it to refine its classification model. [21] use active learning via sequential design for detecting money laundering.

Observed financial transactions can be summarized as a graph with entities (e.g., accounts) as nodes. Graph mining techniques can be used to identify suspicious money flows across the edges of such a graph. [22] propose a new Link Discovery based on Correlation Analysis (LDCA) on timeline data to identify communities in the absence of explicit link information. The correlation between two persons is defined through a correlation function between their financial transaction history vectors. If both are part of the same ML episode, they should exhibit similar financial transaction patterns, and thus, one expects a higher correlation value for them. [23] propose a graph mining method for the detection of subgraphs corresponding to suspicious transaction patterns (e.g., a lattice-like sender-intermediaries-receiver pattern). Their method takes into consideration dependencies between individual transfers that may be indicative of illegal activities; see also [24]. [25] presented a set of coordinated visualization metaphors including heatmap, search by example, keyword graphs, and strings and beads, which are based on identifying specific keywords within the wire transactions. This set of visualizations helps analysts to detect accounts and transactions that exhibit suspicious behaviors. [26] applies statistical outlier detection techniques to detect ML episodes in import and export trades data, since overvaluing imports or undervaluing exports is a common ML method. [27] has developed a set of active agents (Sentinels), along with probabilistic methods, to detect unusual events and entities indicative of ML. [28] present two interesting unsupervised techniques to identify suspicious entities. In peer group analysis, an entity (e.g., account) is selected as a target and is compared with all other entities in the database, and a peer group of entities most similar to the target object is identified. The behavior of the peer group is then summarized at each subsequent time point, and the behavior of the target entity is compared with the summary of its peer group. Those target entities exhibiting behavior most different from their peer group summary behavior are flagged. Break point analysis (BPA) slides a window over the sequence of transactions of an account and uses statistical tests to compare a window with earlier ones to detect any sharp changes in the transaction patterns (e.g., frequency, amounts). [29] developed a cluster-based outlier detection algorithm to identify suspicious ML transaction patterns.

[30] have proposed a supervised decision tree algorithm for ML detection, combined with a privacy preserving strategy (Inner Product Protocol) to protect the identity of the account owners, in case they are not identified as part of suspicious ML. [31] discuss a methodology for AML, in which many steps make use of various data-mining techniques such as outlier detection, link analysis, and community detection.

Given the scarcity of labeled data, generating and using synthetic transactions data that can contain known suspicious patterns is important for validating the ML detection algorithms.

6   Fraud and Money Laundering Detection: A Roadmap

Clearly, much work has been done in developing analytics techniques for the detection of frauds and money laundering. However, there is much more to do. Following are some suggestions about areas where more research is needed.

1. Since fraud detection systems are used by domain experts and business executives, there is a need to develop interactive visual techniques to help users quickly explore large and complex datasets and spot interesting patterns. Support for post-facto evidence gathering is important.
2. IT systems in most domains collect diverse types of data; e.g., insurance claims data contains structured information, textual data (police reports, witness statements, adjustor comments), images (car damage photos), audio and video (phone calls, witnesss interviews), graph data (money flow graphs, stock flow graphs, supplier networks), time-series (timestamped transactions and events) etc. Multimodal data mining needs to be developed that can

mine data of diverse types and seamlessly and systematically integrate the results, using *knowledge fusion techniques* that need to be developed.

3. Evidence, both data-based and human-gathered, plays a crucial role in fraud management processes that happen after a fraud occurrence is detected. Discovering evidence is a slow, effort- and knowledge-intensive human process, which is highly subjective. There is a need to develp a formal theory of various types of evidences and systematic interactive methods to gather, record and combine them, which can then become an integrated part of any fraud control system. The current practices of defining and combining red-green flags also needs formal foundations. All this is important also to control the relatively high false alarm rates seen in practice, which increase the workload on investigation experts and annoy genuine customers.

4. Fraud control and management systems involve complex business processes and skilled personnel. There is a need to interate and align the fraud analytics algorithms with these business processes, so that right insights and patterns are available at specific decision-points. For example, selecting the right recovery strategy for a particular fraud case may benefit from insights obtained by an analysis of past recovery efforts.

5. There is a need to develop techniques for estimating the extent of false negatives (fraud cases that were missed out). No such techniques seem to be available for stock market frauds. Techniques such as anomaly detection and active learning can be combined to probe and discover emerging newer fraud cases that fall outside the "standard" fraud methods.

6. To boost the research and to facilitate comparative studies, there is a need to make available shared public-domain datasets of known fraud cases. For example, so far as we know, there are no shared datasets for stock market frauds. To compensate for the lack or sparsity of known fraud cases, there is a need to develop specialized unsupervised techniques (profiling, fraud *signature* analysis, outlier detection, clustering, community detection, dependency analysis etc.), as well as adapt techniques from semi-supervised learning, deep learning, transfer learning and co-training. Common fraud aspects such as collusion need to be investigated from several perspectives such as game theory and applied psychology.

7. It is clear from practice that fraud detection is domain-knowledge intensive, with consequent dependence on domain experts. Even feature engineering for building effective classifiers needs a number of data elements collected manually by humans; e.g., flags such as Claimant appeared claim-wise, Claimant was uncooperative etc. There is a serious need to understand, formalize, capture, reason with and integrate this domain-knowledge into various analytics algorithms. There is a need to develop a shared, standardized and formal fraud typology across domains. Further, humans have intuitive models of fraud *modus operandis*, based on which we can develop formal knowledge-models for various fraud types. Notations such as rules, graphical models have already been used to build such models. Knowledge representation theories from AI such as modal logics or description logics can be used to build such models. However, since frauds are very much a human activity, there

is a need to enrich the data-driven aspects of such knowledge-models with knowledge from criminology and applied psychology.

8. There are challenges in devising useful measures of the accuracy of fraud detection in operation. Practical "end-to-end" measures for the effectiveness of fraud control processes (conviction rates, recovery ratios) need to be understood formally and related to more standard detection measures like % correct cases and AUROC; e.g., how to design analytics algorithms to detect frauds which maximize the recovery?

9. Emergence of newer fraud types is a constant threat, which needs more research. For example, how can we "anticipate" newer frauds, if new products (e.g., new automobile insurance policies) or newer technologies (e.g., bitcoin) are introduced? This needs a formal modeling of the notions of vulnerability of business products and processes.

7 Conclusions

This tutorial is an introduction to the general area of fraud detection to analytics scientists and professionals. We discussed some analytics techniques used in their detection. We focused specifically on frauds in insurance, in stock markets and on money laundering. We also discussed some open research problems.

References

1. Bolton, R., Hand, D.: Statistical fraud detection: A review (with discussion). Statistical Science 17, 235–255 (2002)
2. Phua, C., Lee, V., Smith, K., Gayler, R.: A comprehensive survey of data mining-based fraud detection research. In: Tech. Report, Clayton School of Information Technology, Monash University, 1–27 (2005)
3. Yue, D., Wu, X., Wang, Y., Li, Y., Chu, C.H.: A review of data mining-based financial fraud detection research. In: Int. Conference on Wireless Communications, Networking and Mobile Computing, pp. 5519–5522 (2007)
4. Mantone, P.S.: Using Analytics to Detect Possible Fraud: Tools and Techniques. Wiley (2013)
5. Spann, D.: Fraud Analytics: Strategies and Methods for Detection and Prevention. Wiley (2013)
6. Subramanian, R.: Bank Fraud: Using Technology to Combat Losses. Wiley and SAS Business Series (2014)
7. Dorrell, D.D., Gadawski, G.A.: Financial Forensics Body of Knowledge. Wiley (2012)
8. Nigrini, M.: Forensic Analytics: Methods and Techniques for Forensic Accounting Investigations. Wiley (2011)
9. Viaene, S., Derrig, R.A., Baesens, B., Dedene, G.: A comparison of state-of-the-art classification techniques for expert automobile insurance claim fraud detection. Journal of Risk and Insurance 69, 373–421 (2002)
10. Tennyson, S., Salsas-Forn, P.: Claims auditing in automobile insurance: Fraud detection and deterrence objectives. Journal of Risk and Insurance 69, 289–308 (2002)
11. Derrig, R.A., Viaene, G.D.S.: A case study of applying boosting naive bayes to claim fraud diagnosis. IEEE Transactions on Knowledge and Data Engineering 16, 612–620 (2004)

12. Phua, C., Alahakoon, D., Lee, V.: Minority report in fraud detection: classification of skewed data. SIGKDD Explorations Newsletter 6, 50–59 (2004)
13. Donoho, S.: Early detection of insider trading in option markets. In: Proc. of the Tenth ACM SIGKDD International Conference on Knowledge Discovery and Data Mining (KDD04), pp. 420–429 (2004)
14. Palshikar, G.K., Apte, M.: Collusion set detection using graph clustering. Data Mining and Knowledge Discovery 16, 135–164 (2008)
15. Huang, M.L., Liang, J., Nguyen, Q.V.: A visualization approach for frauds detection in financial market. In: Proc. 13th Int. Conference on Information Visualisation, pp. 197–202 (2009)
16. Palshikar, G., Apte, M.: Financial security against money laundering: A survey. In: Akhgar, B., Arabnia, H.R. (eds.) Emerging Trends in Information and Communication Technologies Security, ch. 36, pp. 577–590. Elsevier (Morgan Kaufman) (2013)
17. Madinger, J.: Money Laundering: A guide for criminal investigators, 3/e. CRC Press (2012)
18. Truman, E., Reuter, P.: Chasing dirty money: Progress on anti-money laundering. Peterson Institute (2004)
19. Turner, J.E.: Money laundering prevention: Deterring, detecting, and resolving financial fraud. Wiley (2011)
20. Senator, T.E., Goldberg, H.G., Wooton, J., Cottini, M.A., Khan, A.F.U., Klinger, C.D., et al.: The financial crimes enforcement network AI system (FAIS): Identifying potential money laundering from reports of large cash transactions. AI Magazine 16, 21–39 (1995)
21. Deng, X., Joseph, V.R., Sudjianto, A., Wu, J.: Active learning via sequential design with applications to detection of money laundering. Journal of American Statistics Association 104, 969–981 (2009)
22. Zhang, Z., Salerno, J.J., Yu, P.S.: Applying data mining in investigating money laundering crimes. In: Proc. SIGKDD 2003, pp. 747–772 (2003)
23. Michalak, K., Korczak, J.: Graph mining approach to suspicious transaction detection. In: Proc. Federated Conference on Computer Science and Information Systems, pp. 69–75 (2011)
24. Moll, L.: Anti money laundering under real world conditions-Finding relevant patterns. MS thesis, Department of Informatics: University of Zurich (2009)
25. Chang, R., Lee, A., Ghoniem, M., Kosara, R., Ribarsky, W., Yang, J., et al.: Scalable and interactive visual analysis of financial wire transactions for fraud detection. Information Visualization 7, 63–76 (2008)
26. Zdanowicz, J.S.: Detecting money laundering and terrorist financing via data mining. Communications of the ACM 47, 53–55 (2004)
27. Kingdon, J.: AI fights money laundering. IEEE Intellent Systems 19, 87–89 (2004)
28. Bolton, R.J., Hand, D.J., David, J.H.: Unsupervised profiling methods for fraud detection. In: Proc. Credit Scoring and Credit Control VII, pp. 5–7 (2001)
29. Zengan, G.: Application of cluster based local outlier factor algorithm in anti money laundering. In: Proc. Int. Conference on Management and Service Science (MASS 2009), pp. 1–4 (2009)
30. Ju, C., Zheng, L.: Research on suspicious financial transactions recognition based on privacy preserving of classification algorithm. In: Proc. 1st Int. Workshop on Education Technology and Computer Science (ETCS 2009), pp. 525–528 (2009)
31. Gao, Z., Ye, M.: A framework for data mining-based anti-money laundering research. Journal of Money Laundering Control 10, 170–179 (2007)

Job Routing in Multi-site Print Shop Environment

Rahul Paul, Maurya Muniyappa, and Pallavi Manohar

Xerox Research Centre India

Abstract. Xerox is a market leader in managing print services and provides enhanced print-shop productivity through its LDP Lean Document Production solution. Geographically distributed and multi-site service operations present challenges that are quite different from stand-alone print shops. There is a need to address other aspects for the multi-site enterprise such as optimal routing of print jobs and capacity planning. This paper presents a system that can assist the enterprise manager to make efficient job routing decisions within the enterprise. The system described is part of the LDP suite enabling multi-site enterprise print management.

1 Overview

A brief description of the problem is schematically described in Fig. 1. For each print job that arrives through a web interface, the objective is to identify a print shop where the job can be processed such that cost to the enterprise to produce and ship it is minimized while the job is delivered on-time to the customer. A discrete event based simulation framework is developed to compute and display the turnaround time (TAT) for each job (which involves processing time, waiting time, and shipping time) and its cost (which comprises of equipment cost, operator cost, and transit cost). Using automated routing algorithms based on TAT and cost of the jobs at different shops, decisions for routing the incoming jobs are made. Our system can also work in a semi-automated mode where the enterprise manager can tune the parameters of a routing algorithm to perform what-if analysis and make routing decisions. Further, the system can be used in a manual mode where the customers are presented with the shop options along with their cost and TAT and then customers select the shop as per their cost-TAT trade-off preferences.

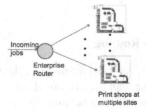

Fig. 1. Job routing in a multi-site print shop environment

S. Srinivasa and S. Mehta (Eds.): BDA 2014, LNCS 8883, pp. 161–164, 2014.

Existing systems rely on the ad-hoc routing decisions made by an enterprise manager or operations manager based on prior experience. Currently, no system exists for automated online routing for print jobs or for assisting the print enterprise manager or the customer in making routing decisions.

2   Enterprise Module of LDP

The enterprise module of LDP comprises of multiple shops in unified modeling environment. The enterprise module enables the user to compare shops with similar functionality for a given set of data or jobs and to provide a detailed view of shop components and parameters (See Fig. 2) that are part of the enterprise.

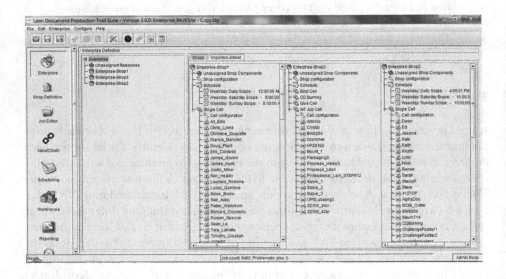

Fig. 2. Enterprise module of LDP software

The system further demonstrates routing of selected print jobs across the shops (See Fig. 3).

2.1   Modes of Enterprise Job Routing

The enterprise module considers various parameters related to cost and current load at the multiple shops, so that the user can get a detailed view of the current enterprise. The report showing the simulation results can be utilized by the customer to decide which shop to route the print job to. The corresponding dashboard provides four views of the analysis to enable routing of jobs within the enterprise

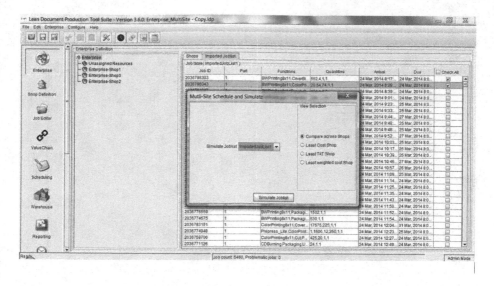

Fig. 3. Functionality showing the ability to view, select and simulate jobs within the enterprise

1. Manual mode
   (a) In this mode, the jobs are processed one at a time and corresponding simulations are run for computing the processing time. If it is a first job assigned to the shop, the Turn Around Time (TAT) and Release time are calculated from the Simulation.
   (b) If some jobs exist within the shop and a job-queue exists, the processing time is calculated from the simulation, and the waiting time for the current job to be scheduled within the shop is computed.
   (c) The waiting time is computed based on current load in that shop and its release time is determined accordingly.
   (d) The TAT which is Waiting Time + Processing Time, for the job for that particular shop is then displayed to the user.
   (e) The above steps are repeated for all the shops available under the enterprise to give the user a detailed view of TAT and cost so that they can compare the performance across all the shops and decide which shop to route the jobs to (See Fig. 4).
2. Automated Mode
   (a) Least cost shop decision
       In this mode, the user is provided an automated solution to make cost-effective assignment of jobs to shops. The system computes and displays the optimal shop assignment to the user. The calculation of TAT is done as described in the manual mode. The cost comprises of processing cost, shipping cost and transit cost. (The demo results are presently based on synthetic data.) In this mode, only the shops that have feasible TAT are considered and the least cost shop is displayed as a decision.

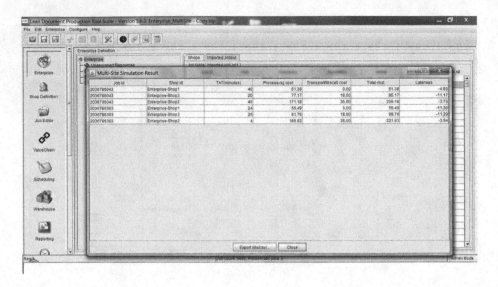

Fig. 4. Display of the results of the multi-site simulation

(b) Least TAT shop decision

This is another view provided to the user where the shop with the least TAT is displayed. Here, cost is not considered to make the routing decision.

3. Semi-automated mode

In this view, a multi-objective decision making framework is presented where the TAT and costs are weighted differently. The weight (w) is the preference given by user or enterprise manager to the TAT and cost. The weight can take a value between 0 to 1. The weighted cost for each shop is calculated as follows.

$$weighted\ cost = w * cost + (1 - w) * TAT \tag{1}$$

The weighting factor input is obtained from the user, the simulations are performed and the least weighted cost shop is displayed.

Energy Data Analytics Towards Energy-Efficient Operations for Industrial and Commercial Consumers

Amarjeet Singh and Vishal Bansal

Zenatix Solutions Pvt. Ltd.
amarjeet.singh@zenatix.com

Abstract. Zenatix Solutions Pvt. Ltd. is an energy data analytics company that helps industrial and commercial entities in optimizing their energy consumption through better understanding of their energy spend. Analysis from the high resolution energy consumption data, collected every few seconds, from smart energy meters installed at sub-load (e.g. Winding Machine, Air Handling Units (AHU), UPS system) level is presented in an easy to comprehend format for the facilities manager helping them identify inefficiencies in their daily operations and reduce their energy consumption. We present the system architecture of how high resolution energy data (and other data that may impact the energy consumption e.g. production in a manufacturing unit or weather for AHU in an IT company) is collected and managed efficiently in a cloud platform. We present several use cases of how the analysis on the collected data helped in identifying the operational inefficiencies across different customers, resulting in significant energy savings.

1 System Architecture

Traditionally, energy consumption data is available to the customers on monthly basis through the electricity bill provided by their utilities. However, such data lacks details that can help in optimizing the overall operations to reduce the energy consumption e.g. performance of different loads at different times of the day, comparison of actual consumption with benchmarks and correlation of energy consumption with related information such as production for manufacturing units and weather for Air Conditioning. Affordable smart meters (electric meters with a communication interface) are changing this scenario by allowing electrical consumers to collect high resolution data (updated every few seconds) from across their facility and better understand their energy consumption as a first step towards optimizing it.

Zenatix leverages this opportunity by working with its customers to first install smart meters and other sensors that provide information relevant for optimizing energy consumption (such as temperature, humidity or light) at appropriate locations. Data from these sensors is collected using a gateway device called zSmart. The zSmart controller has several unique features, including:

1. Interface with smart meters from a variety of vendors such as Schneider Electric, Secure, ElMeasure,Trinity, etc.

S. Srinivasa and S. Mehta (Eds.): BDA 2014, LNCS 8883, pp. 165–168, 2014.

2. Communicate data to a cloud server over Ethernet/WiFi/GSM interface
3. Local storage (capable of storing data for more than a year) to account for intermittent network connectivity
4. Sample meters at different rates within the same loop
5. Secure communication using https
6. Interface with a wide variety of sensors and collect data over diverse interfaces (such as Modbus, 1-wire, I2C, Analog and Digital IOs)
7. Integrate any Building Management System (BMS) using standard BACnet interface for advanced analytics
8. Linux based system allowing for remote access and easy configurability

Data collected from zSmart is pushed to the Zenatix cloud server and is managed in readingDB [1], a database optimized for time-series data. This data can be queried using a simple query language. An analytics layer is built on top of the data storage that also interacts with the frontend to provide useful visualizations. The frontend is responsible for presenting the data in an easy to comprehend manner for the facility manager. Modular architecture for the system allows it to easily scale across distributed machines, for large number of customers and for performing analytics on large volumes of collected data. Figure 1 presents the overall system architecture used for collecting high resolution data from a customer facility. This architecture includes customized enhancements and visualizations built on top of open source sMap architecture [2].

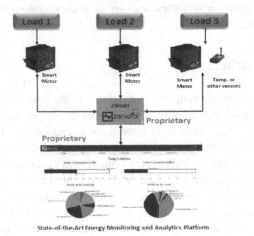

Fig. 1. Zenatix solution architecture for industrial/commercial consumers

2   Use Cases

Current list of customers for Zenatix include college campuses, manufacturing units, IT companies, BPOs and consumers trading on energy exchange. At each of these locations, sampling interval varies from 3 seconds to 60 seconds. Zenatix typically collects 8-10 parameters from each energy meter, resulting in approx. 25k data points from one meter in 1 day. Our biggest deployment is at a university campus with more than 200 energy meters resulting in more than 5 million data points being collected from the campus every day. Table 1 presents use cases for different consumers demonstrating how simple analysis on the collected data helped in identifying inefficient operations and reduce the energy consumption (resulting in a typical payback period of less than one year)

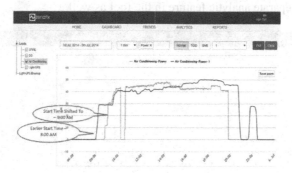

Fig. 2. Zenatix solution helps customers reschedule the start times of AHUs

Table 1. Use cases of energy consumption optimization for Zenatix consumers achieved with the data collected from their facilities

Customer	Zenatix Delivered Value	Customer Benefit
IT Company	Shift the start time of AHUs later by 1 hour without compromising on the comfort. (see Fig. 2)	This resulted in savings of more than Rs. 5 lakhs/annum
IT Company	1. Alerts ensuring AC switch off while DG is operating (as per their policy). 2. Identification of a dead load that shouldn't run at night and weekends. 3. Delayed (optimum) start of ACs.	This resulted in savings of more than 6%.
College Campus	1. Rescheduling of non-critical loads such as ETP and ground water pump to take advantage of the Time of Day (TOD) based pricing. 2. Optimized scheduling of chillers 3. Load optimization on UPS to get maximum efficiency as per the load-efficiency curves	Reduction in monthly energy spend by 10%. See Fig. 3
Manufacturing Unit	The customer is able to monitor the start/stop times of shifts, and optimize consumption during the lunch/tea breaks. (See Fig. 4)	Increased transparency in operations. Monetary benefits are yet to be quantified.
Manufacturing Unit	As a first step, the customer was able to identify and optimize losses due to DG operation after the utility supply is restored.	This action itself resulted in monthly savings of more than Rs. 30,000 (more than 3%).

Figure 2 provides an illustration of savings achieved in the IT company due to shift in AHU start times (also referred to as coasting). Peak power scatter plot (shown in Figure 3) presented through Zenatix platform helps in better understanding of consumption at different times of the day and exploit the TOD pricing to reduce the overall electricity spend. Real time trending (shown in Figure 4) helps in better understanding of energy spend during different break times. This option is currently being automated as a separate analytic feature in the tool as well.

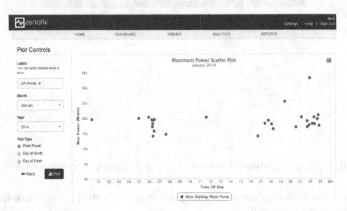

Fig. 3. Peak Power scatter plot helped identify loads to be shifted to off-peaks hours

Fig. 4. Trends Plot helps identifying the exact duration of breaks

References

[1] https://github.com/stevedh/readingdb
[2] Dawson-Haggerty, S., Jiang, X., Tolle, G., Ortiz, J., Culler, D.: sMAP: a simple measurement and actuation profile for physical information. In: Proceedings of the 8th ACM Conference on Embedded Networked Sensor Systems, pp. 197–210. ACM (2010)

Exploring Kaguya Moon Mission Sensor Data by Locating Geographic Entities

Hikaru Suzuki, Eri Suzuki, Wanming Chu, Junya Terazono, and Subhash Bhalla

Database Systems Laboratory, University of Aizu, 965-8580, Japan

Abstract. The Kaguya Lunar probe generated 88 data products using various instruments, during its long mission. These products are based on location and time. These include, Carbon dioxide readings, altitude, images, temperature, and so on. The raw data has been published and is available for download in the raw form. A location name based search facility has been proposed and implemented for exploration and visualization of scientific data for the data products.

Keywords: visualization of scientific data, Management of streamed data archive.

1   Introduction

The Japan Aerospace Exploration Agency (JAXA) launched a lunar orbiter named "Kaguya" (English name: SELENE) [6]. This orbiter observed many kind of moon data for about one year and a half until June 2009. The objectives of the mission were to study the lunar surface environment [5]. The Kaguya orbiter also performed a near moon space study of radio science, plasma, electromagnetic field and high-energy particles through its instruments. Therefore, the data is being analyzed by lunar scientists around the world on a large scale.

2   Kaguya Data Archive

Observational data obtained by Kaguya is processed to reduce noise and arranged to form proper data formats. The data from each instruments is called as "Product". It is produced by an instrument by varying product level or instrument function. These mission products are now available as "Kaguya Data Archive" [3], supported by JAXA. Using the Archive, one can search and download a piece of data items by specifying the search condition such as product ID, latitude/longitude, time range and area (selection). These conditions are given as per the definitions from the metadata (e.g., observation date, instrument name, version number, time or geo-position). The metadata is associated with individual single observed file. It is called "Catalog Information" in the archive. Hence, the detail observation file search with Catalog Information is available in this archive. The Kaguya Data Archive does not have a search function that uses lunar features or location (or object name).

S. Srinivasa and S. Mehta (Eds.): BDA 2014, LNCS 8883, pp. 169–173, 2014.

Similar data archives by NASA provide the "Lunar Mapping and Modeling Project (LMMP)" [7] and the "Lunar Orbital Data Explorer" [10]. Both web sites provide access through the lunar features (e.g., crater, mountain, and sea) and keyword search function as shown in Fig. 1.

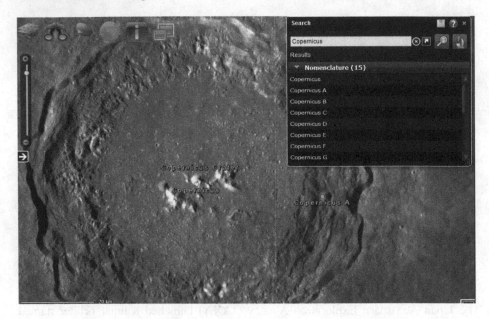

Fig. 1. Screenshot of LMMP with keyword search

In order to support visualization tools for Kaguya data, the basic features for locating data through lunar geographic features and names was considered as essential. Similarly, an easy to use query language interface has been evolved that permits queries.

For this purpose, a lunar location based search system has been developed, called the lunar location search site ``Moon Seeker'' [9]. It can be used for searching by Query-By-Example method using lunar feature, longitude/latitude, diameter, ethnicity, and names of geographic features ( Fig. 2 ). The Moon Seeker enable high-level compound retrieval like ``Find location which is within 100km from previous search result's location'' or ``Find location which is overlapped with previous search result's location''. In addition, we can visually check the retrieved location by use of Google Earth API [1]. This function displays the search result as if those are mapped onto the lunar Geographic Information System (GIS). Thus, as a query language supporting a search system, it enables users to do querying, as compared to the ``location find function'' which come with above mentioned NASA's data archive (LMMP and LODE).

3   Moon Seeker for Kaguya mission

Common specification for location co-ordinates are used by data archives of the NASA orbiters, and the Kaguya orbiter and the proposed lunar location search systems. Moon Seeker has a database of Geographic locations with longitude and latitudes. It was prepared by downloading a lunar geographic database from the NASA data repositories. The database has been utilized to generate the required co-ordinates. The Kaguya's archives deal with latitude/longitude for handling the observation data and lunar locations.

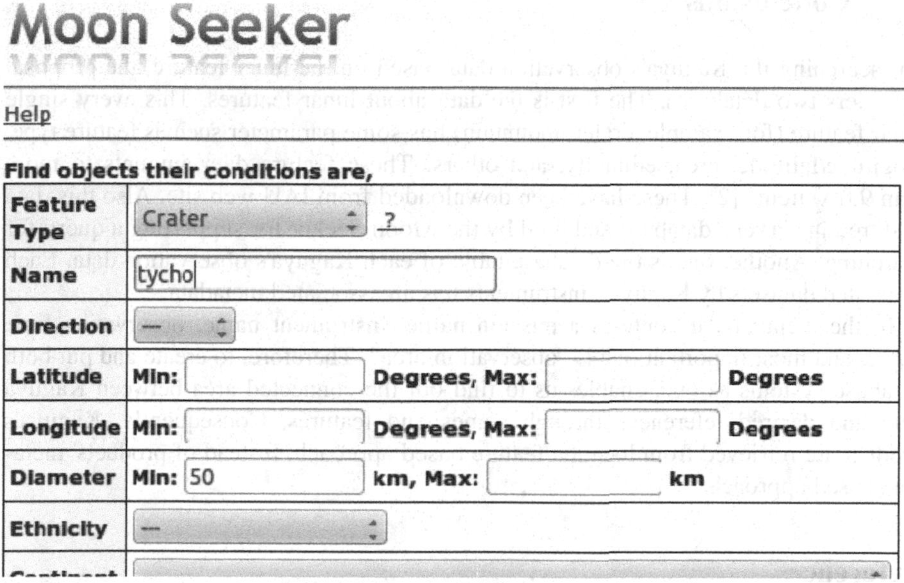

Moon Seeker

Help

Find objects their conditions are,

Feature Type	Crater $\updownarrow$?		
Name	tycho		
Direction	— $\updownarrow$		
Latitude	**Min:**	**Degrees, Max:**	**Degrees**
Longitude	**Min:**	**Degrees, Max:**	**Degrees**
Diameter	**Min:** 50	**km, Max:**	**km**
Ethnicity	— $\updownarrow$		

Fig. 2. Search condition page of the Moon Seeker

Therefore, by extracting the Catalog Information from each Kaguya's observation data, it enables displaying the Kaguya's observed data. In this study, we combine the Moon Seeker and the Catalog Information of Kaguya's data to make the new Kaguya archive query and search system.This system aims to support the researcher who study the Kaguya data using products from various instrument settings. The purpose is to improve access to Kaguya data.

4   Method

To develop the mashup service with ``Kaguya Data Archive'' with ``Moon Seeker'', we combine two systems. The lunar explorer ``Kaguya'' had been equipped with 15 instruments. A product is produced by an individual single instrument. The number of Products which were created by Kaguya mission are of 88 types [4]. In these individual Products, there are a large number of archived files. These have a different time range and observation area. Each archived file consist of image or numerical primary data with metadata (named "Catalog Information") and thumbnail image.

As can be seen above, the structure of Kaguya data is composed of an instrument by its identity, which has a few Products. A Product has many archived files which is lumped primary data and ancillary data together. This structure is known as Planetary Data System (PDS) [8].PDS was invented by NASA, and this is very popular and widely used format in any planetary science projects. Kaguya's data is also created according to the same PDS format.

5   Conclusions

For searching the Kaguya's observation data based on the lunar feature, the proposal considers two databases. The first is the data about lunar features. This every single lunar feature (for example, crater, mountain) has some parameter such as feature type, longitude/latitude, area, ethnicity, and others. These feature data amounts to more than 9,000 items [2]. These have been downloaded from IAU web site. Also this data is stored in server's database and used by the Moon Seeker for supporting a query for searching. Another one is the database table of each Kaguya's observation data. Each generated datasets by Kaguya's instruments has an associated metadata.

In the metadata, it contains a mission name, instrument name, observation time range, and most important one is "observation area". Therefore, to create and put both database systems as one, enables us to find out the connected area between Kaguya data and desired references through names and features. Consequently, Kaguya's product are retrieved from location/feature based approach, instead of products' metadata based approach.

References

1. Google Inc. Google Earth API - Sky, Mars, and Moon, https://developers.google.com/earth/documentation/sky_mars_moon (accessed January 27, 2014)
2. International Astronomical Union (IAU). Gazetteer of Planetary Nomenclature - The Moon, http://planetarynames.wr.usgs.gov/SearchResults?target=MOON (accessed January 27, 2014)
3. Japan Aerospace Exploration Agency (JAXA). Kaguya Data Archive, https://l2db.selene.darts.isas.jaxa.jp/ (accessed January 27, 2014)
4. Japan Aerospace Exploration Agency (JAXA). Kaguya product list for public, http://l2db.selene.darts.isas.jaxa.jp/help/en/KAGUYA_product_list_public_en.pdf
5. Japan Aerospace Exploration Agency (JAXA). KAGUYA (SELENE) - Mission Profile, http://www.kaguya.jaxa.jp/en/profile/index.htm (accessed January 27, 2014)
6. Japan Aerospace Exploration Agency (JAXA). Kaguya (SELENE, Selenological and Engineering Explorer), http://www.kaguya.jaxa.jp/index_e.htm (accessed January 27, 2014)

7. National Aeronautics and Space Administration (NASA). Lunar Mapping and Modeling Project (LMMP), http://pub.lmmp.nasa.gov/LMMPUI/LMMP_CLIENT/LMMP.html (accessed January 27, 2014)
8. National Aeronautics and Space Administration (NASA). The Structure of PDS Data Sets, http://pds.nasa.gov/data/data-structure.shtml (accessed January 27, 2014)
9. Tsunokake, T.: Moon Seeker: Search System for a Lunar Geographic Information System with a Query-By-Object Interface. Master's thesis, The University of Aizu (March 2010), http://datadb04:8090/moon_seeker
10. Washington University in St. Louis.Lunar Orbital Data Explorer, http://ode.rsl.wustl.edu/moon/ (accessed January 27, 2014)

NewsInstaMiner: Enriching News Article Using Instagram

Mohit Thakur, Dhaval Patel, Shrikant Kumar, and Jayendra Barua

Department of Computer Science and Engineering,
Indian Institute of Technology Roorkee
Uttrakhand -247667 India
{mohthuec,patelfec,skantdec,jbarudec}@iitr.ac.in

Abstract. Instagram is one of the largest and most popular image sharing web-site. Millions of users share their real life events with the outside world by up-loading images on Instagram. Most of these images are annotated with tags and captions which are high level descriptions of these images. So if a news event occurs, users might have captured the relevant images and share them on Insta-gram. We have observed that many images on Instagram are relevant to the popular News events. So we can recommend such images for the news articles related to the event. To the best of our knowledge, no existing work has been done yet to recommend images particularly for news articles. In this paper, we have developed an image recommendation technique namely NewsInstaMiner. This technique analyzes news articles to extract the high level concepts and based on these concepts it recommends images from Instagram. We have im-plemented NewsInstaMiner and used it to demonstrate the effectiveness of our proposed technique.

Keywords: Instagram; image recommendation, concept extraction.

1   Introduction

Recent advancement in Internet and Communication Technology (ICT) has revolutio-nized many fields including news media. Traditionally, news sources use print media to disseminate information. However, more and more people use a smart phone that instantly provides "any time anywhere" access to news. Adapting the new technology, now majority of the news sources publish news through online news websites, for example, Times of India, The Hindu etc. Publishing news online not only provides instant access to the news, but also encourages users to search for other relevant in-formation. Generally article published on news websites contains four fields: 1) News Headline, 2) Article text, 3) Relevant images and 4) News publication date. For example, in Figure 1, a News article is shown with the headline "*37 killed as Israel bombs Gaza, Hamas fires rockets*". However, sometimes images may or may not present in News articles. Images make article more Interesting and informative, so news readers always expect some visual information related to the news events. Such images may be available on other sources such as a social networking and image

S. Srinivasa and S. Mehta (Eds.): BDA 2014, LNCS 8883, pp. 174–188, 2014.

sharing websites such as Facebook[1], Instagram[2], Flickr[3], Picasa[4]. But these images are not linked with the online news article. To address this problem, we develop a new approach to recommend News article related images, namely NewsInstaMiner.

Among all of the aforementioned images sharing websites, Instagram is one of the most popular because of its high user engagement ratio. According to Cnet [1], Instagram has more than 200 million active users. As of December 2014, around 16 billion images have been posted to the platform, with an average of 55 million daily [5]. Being one of the most popular and heavily used website, Instagram can provide images related to particular News event or theme. As the user base of Instagram will continue to grow in volume, we have decided to make the most out of this platform for our NewsInstaMiner. In summary, we are using Instagram for fetching event related images. As the image sharing using smart phones is quite common these days, people often capture the images of the events playing around them and update them on image sharing websites. While uploading images to Instagram, users generally insert tags and captions with the image to make it more searchable. Here, tags provide contextual information to the image. In particular, each uploaded image has three information associated with it. These are Time-Date, Caption, and Hash-tags. Caption is the high level description of the image and Hash-Tags generally preceded by a hash sign (#) which oftenly used on social media sites such as Twitter to identify messages on a specific topic.

There have been instances where news source have used Instagram for image recommendation, for example, when CNN used Instagram images from public for inauguration of President Obama [6] and when Electiongram was introduced to cover live events during elections [7]. But these techniques are manual and only used for specific News events. On the other hand, researches [2,3,4] have mainly focused on the image tag recommendation. Some other papers [8,9,10,11,12,13,14] have also done significant work on multimedia information retrieval. We also found that there is no such existing work that enriches news articles using images obtained from different sources.

Our motivation behind creating NewInstaMiner is to provide more and more visual content for the textual news articles. Occurrence of an Interesting news event increases the curiosity of the outside world to know more about it. To quench the curiosity, people read News articles and browse images of the event. Our objective is to provide all this information on a single platform. So, In this paper, we propose a novel technique to enrich News article with relevant images using extracted concepts from News article and Instagram Hashtags. Note that, captions and tags of images play an important role in the retrieval of relevant images from Instagram. An example of our approach is shown in Figure 1. In the given example, the highlighted word "*Gaza*" in News headline is a concept which is used to fetch the image from Instagram. As it's clear from the given example that Instagram contains the images related to the most popular events. So, we can always extract such images from the Instagram which are relevant to News article.

---

[1] https://www.facebook.com/
[2] http://instagram.com/
[3] https://www.flickr.com/
[4] http://picasa.google.co.in/

37 killed as Israel bombs Gaza, Hamas fires rockets

Updated: Wednesday, July 9, 2014, 16:47 [IST]

Gaza/Jerusalem, July 9: In the worst flare-up in two years, at least 37 Palestinians have been killed as the Israeli military today stepped up its offensive pounding Hamas targets in Gaza while almost the whole of Israel came under rocket fire from Palestinian militants.

Hamas-ruled Gaza witnessed its bloodiest day since November, 2012 with 24 deaths since last evening, including women and children.

More than hundred Palestinians were also injured in Israeli Air Force strikes across the coastal strip which targeted 160 sites, including homes of eight senior Hamas operatives, Muhammad Sinwar and Ra'ed Atar, as efforts to stop rocket attacks on Israel's southern areas intensified. The Israel Defence Forces (IDF) said their houses served as Hamas command centres and were used in coordinating the rocket fire on Israel's south.

Israeli warplanes also bombed the home of a senior Islamic Jihad militant in Beit Hanoun post-midnight, killing him and five members of his family. A Gaza health ministry spokesman confirmed that the attack killed Hafiz Hamad and five members of his family.

Recommended Image from Instagram

Associated Hash-tags

#play #palestine #Germany **#gaza** #france #fun #argentina #america #anime #arabs #cool #cute #italy #barcelona #best #romance #sorry #selfi #sexyy #sad #love #lol

Fig. 1. A News article and the corresponding recommended Instagram image based on matched concept *"Gaza"*

The working of NewsInstaMiner goes as follows: (1) NewsInstaMiner periodically fetches the News articles from News Websites. (2) For each article, it extracts news concepts using headline and article text. (3) The extracted concepts are used to retrieve images from Instagram. (4) Finally, it recommends top-K most relevant images to the user based on the similarity in Instagram image Hash-tags and extracted concepts   from News article.

There are two main technical challenges in automating NewsInstaMiner method. Firstly, News article headlines doesn't contain enough keyword tags to perform image query on instagram.   Secondly, the image query on the instagram returns a huge number of images (Shown in Table 1), so finding the most relevant images is a challenging task. In summary, our NewsInstaMiner addresses two main research issues:

Table 1. Number of images fetched from Instagram corresponding to News Headlines

News Headlines	Total images
37 killed as Israel bombs Gaza Hamas fires Rockets	1334
Bullet train concept impressive but not feasible in India:Nitish Kumar	452
BJP demands judicial probe into Rohtas police firing	706
FirstLook:Parineeti-Aditya starrer Daawat-e-Ishq	369
Live Cricket Score India vd England First Test Day 1: India steady after early Dhawan loss against England	536

Research Issue 1. Given a News article A for News event E, extract the concept set $C = \{c_1, c_2, ..., c_x\}$ from News content, which completely describes new event E in terms   of keywords.

Research Issue 2. Next, we retrieve a set of images m_i from Instagram corresponding to each concept $c_i \in C$. So, finally for concept set C, we have   image set $M=\{m_1 \cup m_2 \cup...\cup m_x\}$. Then, identify top-K images from set M such that, they are most relevant to E.

Our experimental results show the efficiency of the proposed technique.

The remaining paper is organized as follows. Section 2 presents system overview of our NewsInstaMiner and describes all the proposed methods. Section 3 presents the experimental evaluation. Section 4 discusses the conclusion and future work.

2   NewsInstaMiner Overview

NewsInstaMiner uses a three phase approach. Each phase has its significant role. Figure 2 diagrammatically illustrates all three phases. In first phase NewsInstaMiner periodically crawl News sources and extract News articles, next it extracts the headline, article text and Date information from news articles. All these three information is input to phase 2 in which "*concepts*" are discovered for each article using this information. In the last phase, we retrieve images from Instagram using discovered "*concepts*" and rank them based on the calculated relevancy with a news article.

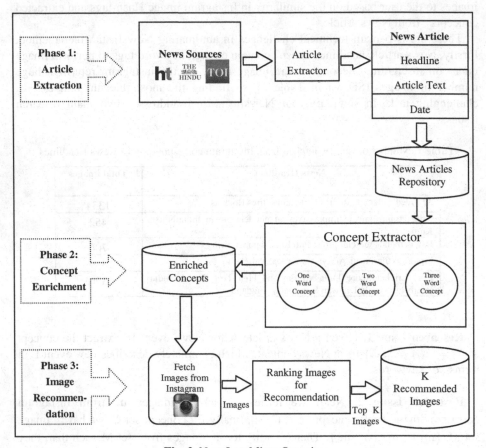

Fig. 2. NewsInstaMiner Overview

All the three phases are described as follows:

2.1   Phase 1: Article Extractor

Since, we have limited no of news sources, so we manually learn the News article web page structure of every news website (see Table 2). For e.g. The Hindu article web page always publishes article text in <div> html tag which has attribute class = "article-text". In the same manner we have defined article extraction rules for every news website. Using these rules we extract News headlines, article text and date of publishing from news article web page.

Table 2. Table shows the recognized HMTL tag structure used for article extraction

News Sources	Headline Tags	Article Text Tags
Times of India	<title>	<div id="artext1" class="Normal showpage1">
The Hindu	<title>	<div class="article-text">
CNN	<title>	<div class="cnn_strycntntlft">
IBN live	<title>	<div id="atxt_box" class="fright">
Hindustan Times	<title>	<div id="ctl00_ContentPlaceHolder1_HTStoryPageControl_strytext" class="sty_txt">

2.2   Phase 2: Concept Extractor and Enrichment

The input for this phase is the headline and the article text of the news article web page under consideration. Our first step is the filtering of article text and headline for concept extraction.

Concept Enrichment: We have used a repository for stop-words. Since these stop-words have no importance to the headline and concept. So headline and article are first filtered to remove all the stopwords. Next, we process the headline to extract the concepts from it. For each word in the filtered headline we first searched it in the article text. For each occurrence of headline keyword in article text, we extract the immediate left and right word of the headline keyword. Thus, we create a headline keyword set including above extracted words with headline keywords. These words may contain some event/action associated with the entity. In this way it enriches the event description in terms of keywords. The example given below illustrates Concept enrichment process.

Table 3. Table shows a snippet taken from a News article

Headline- 37 killed as Israel bombs Gaza hamas fires rockets
Article- Israel readies for ground offensive against Hamas, 12 killed in worst flare-up in two years, at least 37 palestinians have been killed as the Israeli military Wednesday stepped up its offensive pounding Hamas....

The Figure 3 shows the result of Step 1. After removing stop words the headlines becomes "37 killed Israel bombs Gaza hamas fires rockets". In Figure 3 root node shows the headline keyword and their child node shows the immediate left and right words of headline keywords inside the article text. The number of child nodes of a particular node reflects the popularity of that word in the article.

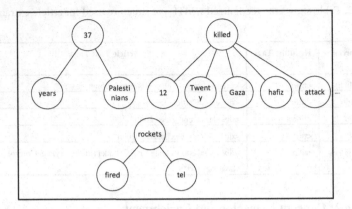

Fig. 3. Concepts after first iteration

We again repeat the concept enrichment process by adding extracted words from the text into the extracted headline keyword set. Figure 4 shows the result of second iteration of Concept Enrichment in the form of tree structures. Each node in the tree represents one word concepts. Each path from one node to another in the trees represent a concept. The number of nodes in the path specifies that it is 1-word, 2-word or 3-word concept. It's clear from the Figure 4 that we got a new node name *"Tel aviv"* which was not mentioned in the graph previously.

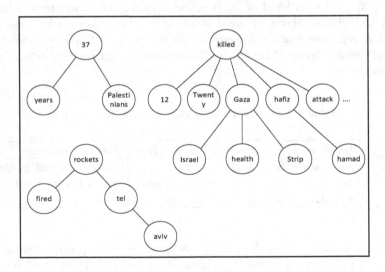

Fig. 4. Concept enrichment after second iteration

In particular, each node represents one word concepts. We will be using our graph structure to make One word concepts, Two word concepts and Three word concepts. We will put all the unique words in One word Category. For above example 64 words were put in that category. We put combination of two word concepts from respective structure in our two word category. For the above example, there were 72 words in this category & finally all the three word concepts in our Three word category having 22 words from example being mentioned above. These concepts will further be filtered out by Instagram in the phase 3. Figure 5 represents our concepts repository.

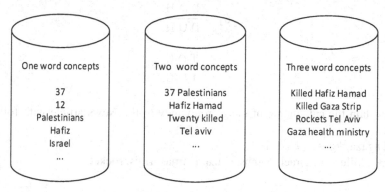

Fig. 5. Concepts repository

2.3   Phase 3: Recommending Images

In this phase, we use the extracted concept from an article to crawl the images from the Instagram. First, we filter out all keywords which are not used as tag in Instagram. We will use all the concepts and check whether they have been used in Instagram history. We get the image count corresponding to each keyword using the API provided by Instagram. For Example, the image count returned by Instagram corresponding to tag "Gaza-Family" is 0. That means there are no images available corresponding to this tag on Instagram.

In our example, we found that only 1 one word concept have been never used, that is "Strip". For Two word Concepts only 18 out of 72 were found to be used as a tag. For three word concepts only one out of 22 found its meaning in Instagram that is IsraelDefenceForces. We can safely assume that after this step only words which have some meaning are defined as concepts and will be used to extract relevant images.

After extracting a set of images for a particular headline keyword set. Our next step is to find relevant images for headline of news article. In this paper, using the properties such as tag and caption we find a relevancy score. The user will generally describe an image using important keywords which are usually found in headline so we compare caption with the headline. Since we are using concepts as tags to extract images, it is a common approach to compare tags with one word concepts. To calculate the relevancy score, we used Jaccard similarity.

T = Set of Tags   used by a user for image
D = Set of Caption/Description of an image

H = Set of keywords in Headline used to extract images
C = Set of Headline keywords concepts ∪ Set of concepts extracted from Article text
G_1 = Caption Headline Jaccard score
G_2 = Tag Concept Jaccard score

$$G_1 = \frac{D \cap H}{D \cup H}$$

$$G_2 = \frac{T \cap C}{T \cup C}$$

We explain the calculation of image relevancy to the News article with following example:
D = {#gaza, #gazze, #israel}
H= {37, killed, as, Israel, bombs, Gaza, Hamas, fires, rockets}

Here G_1 comes out to be 0.1818. As "gaza" and "Israel" are common, so, D∩H = 2 and Total length ("as" is a stop word) i.e. D∪H = 11, G_1 = 2/11=0.1818

T = {gaza, israel, gazze}
C = {Israel, hafiz, hamad, hamas, aviv, muhammad, gaza, attack, killed, warplanes, militants, targets, offensive, defense, forces, tel, 37, Twenty, rockets, islamic, died, army....}. There are total 67 concepts.

Here G_2 comes out to be 0.028. 'gaza' and 'Israel' are common. So
T ∩ C = 2 and Total   length is 70 i.e. T∪C = 70, G_2 = 2/70 = 0.028

3   Experimental Evaluation

In order to evaluate out NewsInstaMiner we implemented a prototype system. Our prototype system displays the user list of all the headlines that have been processed. User has to click the news or headline of his choice and he will get the following details such as date, article, list of images recommended by description similarity and list of images recommended by tag similarity. Figure 6 shows the interface of our system.

In this section we will go through one particular example. We will go though the entire recommendation process. Suppose one particular news headline from news website Hindustan Times is "Germany hit the highs under Loew humiliate Brazil 7-1 on home soil" published on "2014-07-09 18:47:01.0".

News Headline Links	Article Text of Selected Headline	Recommended Images
1. Bullet train concept impressive but not feasible in India: Nitish Kumar	**Germany hit the highs under Loew humiliate Brazil 7-1 on home soil**	Caption: MINEIRAÃ‡O!!! #Scolari #Loew #Brazil #Germany #worldcup #worldcup2014 Tags: brazil,worldcup2014,germany,worldcup,scolari,loew **Jaccard Caption: 0.1875**
2. 37 killed as Israel bombs Gaza Hamas fires rockets	*2014-07-09 18:47:01.0*	Caption: best smile #Loew#germany Tags: germany,loew **Jaccard Caption: 0.142857**
3. Singaporean of Indian-origin fighting in Syrian civil war	*Brazil's humiliating exit from the World Cup is as close as it gets to a national trauma in a football-mad nation whose identity is*	Caption: #BrazilvsGermany#Brazil#Germany Tags: brazil,germany,brazilvsgermany **Jaccard Caption: 0.142857**
4. BJP demands judicial probe into Rohtas police firing	*closely associated with its team. A Brazilian soccer fan reacts as she watches the 2014 World Cup semi-*	Caption: #LMAO this is exactly how the between Brazil vs Germany played. #brazil Tags: brazil,worldcupsemifinals,lmao,germany7 ,worldcup,lol,germanysuperior,germany,humiliate,cutass **Jaccard Tags: 0.142857**
5. Govt rejects WHO study report on pollution level in India	*final. REUTERS The country of 200 million people had hoped to win the title at home and erase*	Caption: #Loew#germany Stately man Tags: germany,loew **Jaccard Tags: 0.133333**
6. Germany hit the highs under Loew humiliate Brazil 7-1 on home soil	*memories of its 1950 World Cup defeat to Uruguay in Rio de Janeiro, the notorious "Maracana-zo" at the Maracana Stadium. Read:...*	Caption: #à¸ˆà¤«æ¼«ç"» #ç"»ç"» #loew #germany #coach #comic #è§Œ™" #è¸...ì ¼#ᵇ"ë"œì»µ Tags: coach,ç"»ç"»,ë¸...ì ¼#§Œ™", germany,ᵇ"ë"œì»µ,æ¼«ç"»,comic,à¸ˆà¤«,loew **Jaccard Tags: 0.133333**

Fig. 6. Snapshot of NewsInstaMiner GUI.

3.1   NewsInstaMiner Output for Article Extraction

We extracted the article text corresponding to the news headline "*Germany hits highs Loew humiliate Brazil 7-1 home soil* " as shown in Figure 7. The article consists 689 words.

3.2   NewsInstaMiner Output for Concept Extraction and Enrichment

Headline- Germany hits highs Loew humiliate Brazil 7-1 home soil
Article- Brazil's humiliating exit World Cup close national trauma....
The length of the article was reduced from 689 words to 368 words.
The result of phase 2 after enrichment is displayed below. No three word concept was found. Table 4 represents one word concepts and Table 5 represents two word concepts.

Table 4. Table shows the 1-word concept extracted from News Article with Headline " *Germany hits highs Loew humiliate Brazil 7-1 home soil* "

Trengrouse	suffered	victories	erase
Mineirao	isolation	seventh	title
Happened	1950	Pedro	falling
Stadium	tears	Read	7
Win	gift	Germany	Brazil
World	1	Today	home

Table 5. Table shows the 2-word concept extracted from News Article with Headline *"Germany hits highs Loew humiliate Brazil 7-1 home soil "*

Brazil-suffered	Brazil-Pedro	Brazil-tears
1950-World	1-Germany	Mineirao-Stadium
Brazil-win	71	

3.3   NewsInstaMiner Output for Recommending Images

In this section we present two instances of NewsInstaMiner output for image recommendation.

Example 1: Figure 7 represents the headline along with the partial article text displayed corresponding to News headline *"Germany hits highs Loew humiliate Brazil 7-1 home soil"* in our system. Figure 8 displays images based on description similarity score. Figure 9 shows images   on the basis of tag similarity score.

Example 2: Figure 10 represents the headline along with the partial article text displayed corresponding to News headline *"Sidharth, Alia, Malaika watch Salman, Jacqueline's 'Kick' along with the Khan-daan"* in our system. Figure 11 displays images based on description similarity score. Figure 12 shows images   on the basis of tag similarity score.

Germany hit the highs under Loew humiliate Brazil 7-1 on home soil
2014-07-09 18:47:01.0
Brazil's humiliating exit from the World Cup is as close as it gets to a national trauma in a football-mad nation whose identity is closely associated with its team. A Brazilian soccer fan reacts as she watches the 2014 World Cup semi- final. REUTERS The country of 200 million people had hoped to win the title at home and erase memories of its 1950 World Cup defeat to Uruguay in Rio de Janeiro, the notorious "Maracanazo" at the Maracana Stadium. Read: The match as it happened But instead Brazil suffered the worst defeat in its 100-year foot-balling history, falling 7-1 to Germany at the Mineirao Stadium in the southeastern city of Belo Horizonte - now the infamous "Mineirazo". "It looked like a game between adults and children," wrote the prominent sports analyst Juca Kfouri on his blog. "Brazilian football has never experienced such humiliation." The country's newspapers called it the biggest disgrace in the team's history, with globoesporte.com dubbing it the "Shame of Shames". But other analysts said the Mineirazo could not compare to the Maracanazo. "In 1950 we felt we had an unbeatable team and losing at the Maracana was unthinkable," Michel Castellar, an analyst at the sports daily Lance, told AFP. Read: We're here to be world champions, say focused Germans "This time, we knew that we had a team with a lot of flaws and that maybe they would not reach the final. Was it a national humiliation? Yes, because of the number of goals. But it wasn't a new Maracanazo," he said. Successful cup In 1950, Brazil had yet to win the World Cup and getting the title would have put the developing country on the map.

Fig. 7. Snapshot of extracted Headline and article text

	Caption: MINEIRAÃ‡O!!! #Scolari #Loew #Brazil #Germany #worldcup #worldcup2014 Tags: bra-zil,worldcup2014,germany,worldcup,scolari,loew **Jaccard Caption: 0.1875**
	Caption: best smile #Loew#germany Tags: germany,loew **Jaccard Caption: 0.142857**
	Caption: #BrazilvsGermany#Brazil#Germany Tags: brazil,germany,brazilvsgermany **Jaccard Caption: 0.142857**

Fig. 8. Snapshot of recommended images from Instagram by NewsInstaMiner with relevancy score based on Description similarity

	Caption: #LMAO this is exactly how the game between Brazil vs Germany played. #brazil #germany #germanysuperior #germany7 #worldcup #worldcup-semifinals #cutass #humiliate #lol Tags: bra-zil,worldcupsemifinals,lmao,germany7,worldcup,lol,germanysuperior,germany,humiliate,cutass **Jaccard Tags: 0.142857**
	Caption: #Loew#germany Stately man Tags: germany,loew **Jaccard Tags: 0.133333**
	Caption: #å‹’å¤« #æ¼«ç”» #ç”»ç”» #loew #germany #coach #comic #ë§Œí™” #ë …ì ¼ #ì›”ë“œì»µ Tags: coach,ç”»ç”»,ë …ì ¼,ë§Œí™",germany,ì›"ë"œì»µ,æ¼«ç"»,comic,å‹’å¤«,loew **Jaccard Tags: 0.133333**

Fig. 9. Snapshot of recommended images from Instagram by NewsInstaMiner with relevancy score based on Tags similarity

Sidharth Alia Malaika watch Salman Jacqueline s Kick along with the Khan-daan
2014-07-25 11:47:01.0
entertainment Sidharth, Alia, Malaika watch Salman, Jacqueline's 'Kick' along with the Khan-daan A special screening was held on Thursday (July 24) of Salman Khan much-awaited action flick 'Kick' that saw the likes of actors Aditya Roy Kapur, Sidharth Malhotra, Alia Bhatt and Nawazuddin Siddiqui. Salman Khan's family including Sohail, Seema, Alvira and Malaika were also present. Kick's lead actress Jacqueline Fernandez was clearly in her element as she made an entrance. ...

Fig. 10. Snapshot of extracted Headline and article text

Caption: Sidharth, Alia and Aditya with Salman at the screening of KICK!! #adityaroyka-poor#sidharthmalhotra#aliabhatt#salmankhan#screening#kick#movie#bollywood#jacquelinefernandez#released#AMAZING
Tags: adityaroyka-poor,aliabhatt,jacquelinefernandez,movie,amazing,released,salmankhan,screening,bollywood,kick,sidharthmalhotra
Jaccard Caption: 0.24

Caption: An evening with the beautiful #Jacqueline-Fernandez #Kick
Tags: jacquelinefernandez,kick
Jaccard Caption: 0.2

Caption: Sidharth, Alia, Aditya and Salman at the premier of 'Kick' ðŸ˜~ #salmankhan #kick #aliabhatt #adityaroykapur #sidharthmalhotra #hottie #perf #indian #bollywood
Tags: aliabhatt,salmankhan,indian,sidharthmalhotra,hottie,adityaroykapur,bollywood,kick,perf
Jaccard Caption: 0.2

Fig. 11. Snapshot of recommended images from Instagram by NewsInstaMiner with relevancy score based on Description similarity

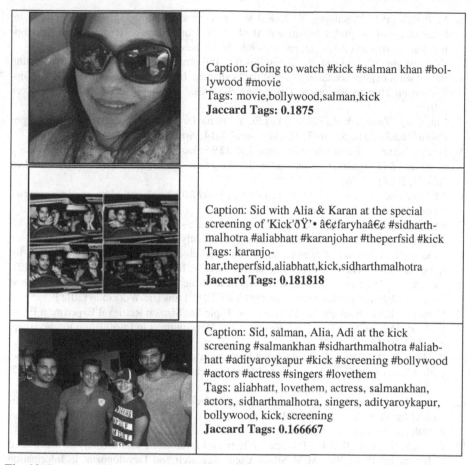

Fig. 12. Snapshot of recommended images from Instagram by NewsInstaMiner with relevancy score based on Tags similarity

4   Conclusion

Image recommendation system called NewsInstaMiner achieves an important task of suggesting relevant images to the user. Instagram can be proved as a useful source for News agencies to gather images to enrich their News stories. The system can automatically gather relevant images for popular News articles. Our experiments show that NewsInstaMiner is promising for news which are popular. In the future, we aim to investigate in more techniques to enrich our concept, so that more relevant images can be gathered.

References

1. http://www.cnet.com/news/instagram-passes-200m-active-users/ (accessed on July 31, 2014)

2. McParlane, P.J., Moshfeghi, Y., Jose, J.M.: On contextual photo tag recommendation. In: Proceedings of the 36th International ACM SIGIR Conference on Research and Development in Information Retrieval, pp. 965–968. ACM, New York (2013)
3. Weilenmann, A., Hillman, T., Jungselius, B.: Instagram at the museum: communicating the museum experience through social photo sharing. In: Proceedings of the SIGCHI Conference on Human Factors in Computing Systems., pp. 1843–1852. ACM, New York (2013)
4. Liu, G.-Q., Zhang, Y.-J., Fu, Y.-M., Liu, Y.: Behavior Identification Based on Geotagged Photo Data Set. The Scientific World Journal 2014, Article ID 616030, 6 pages (2014)
5. http://www.fastcompany.com/3029395/bottom-line/how-the-most-successful-brands-dominate-instagram-and-you-can-too (accessed on July 31, 2014)
6. http://edition.cnn.com/interactive/2013/01/politics/inauguration-ireport/ (accessed on July 31, 2014)
7. http://gigaom.com/2012/11/03/how-nbc-is-using-instagram-to-report-the-2012-election/ (accessed on July 31, 2014)
8. Tsikrika, T., Kludas, J., Popescu, A.: Building Reliable and Reusable Test Collections for Image Retrieval: The Wikipedia Task at ImageCLEF. IEEE MultiMedia 19, 24–33 (2012)
9. Tsikrika, T., Popescu, A., Kludas, J.: Overview of the Wikipedia Image Retrieval Task at Image- CLEF. In: Working Notes for the CLEF 2011 Labs and Workshop (2011)
10. Voorhees, E.M., Buckley, C.: The Effect of Topic Set Size on Retrieval Experiment Error. In: Proc. 25th ACM SIGIR Conf. Research and Development in Information Retrieval, pp. 316–323. ACM Press (2002)
11. Buckley, C., Voorhees, E.M.: Retrieval Evaluation with Incomplete Information. In: Proc. 27th ACM SIGIR Conf. Research and Development in Information Retrieval, pp. 25–32. ACM Press (2004)
12. Zobel, J.: How Reliable Are the Results of Large- Scale Information Retrieval Experiments? In: Proc. 21st ACM SIGIR Conf. Research and Development in Information Retrieval, pp. 307–314. ACM Press (1998)
13. Büttcher, S., et al.: Reliable Information Retrieval Evaluation with Incomplete and Biased JudgmentsIn: Proc. 30th ACM SIGIR Conf. Research and Development in Information Retrieval, pp. 63–70. ACM Press (2007)
14. Müller, H., et al. (eds.): ImageCLEF: Experimental Evaluation in Visual Information Retrieval. Springer (2010)

SARROD: SPARQL Analyzer and Reordering for Runtime Optimization on Big Data

Nishtha Tripathi[1] and Subhasis Banerjee[2]

[1] LNM Institute of Information Technology
nishtha.lnmiit@gmail.com
[2] IIIT-Delhi
subhasis@iiitd.ac.in

Abstract. Resource Description Framework has been widely adopted for representing web resources and structured graph data model. Evolution of Big Data poses challenges in processing these graphs in terms of scalability as the size of the graphs may become enormously big. Tremendous growth of the data size (and in turn the graph size) will have an impounding effect on the execution times of the queries on the graph. Although big data platforms such as Hadoop can mitigate the problem, the query ordering and data flow between the constraints presents opportunities for further optimization. SPARQL, a widely used RDF query language suffers from the similar bottleneck for large graphs. There is hardly any established method to generate all equivalent reordering for a SPARQL query containing joins, outer joins, and group by aggregations. In this paper, we propose a query reordering algorithm viz., **SARROD** that leverages the property of the graphs that are simple to compute yet powerful for run time optimization. Experimental results show that **SARROD** reduces response time for SPARQL queries when executed over SHARD graph-store (triple-store) built on the Hadoop implementation of MapReduce by an order of 12% compared to non-ordered sequence.

1 Introduction

Semantic web technologies will continue to augment human reasoning. Proliferation of data centric computation and generation of large data have yielded the challenge to process large RDF graphs. As volume of data continues to grow, real time processing and querying of data will continue to be a bottleneck for the performance of applications. While Big Data platforms such as Hadoop [1] offers immense benefit in completely parallelizing the execution of the queries, there is significant scope to achieve better performance by optimizing appropriate order of the clauses. Reordering of SQL queries have been explored in several prior work to achieve execution time performance gains in relational databases. We borrow the similar idea from the domain of relational databases and focus on RDF graphs to improve the performance by reordering the queries (in SPARQL).

Technically, RDF databases are formed by triples (*subject, predicate, object*) where each triple expresses binary relation between *subject* and *object* through *predicate*. SPARQL being a declarative query language on RDF, it also comes with the basic

S. Srinivasa and S. Mehta (Eds.): BDA 2014, LNCS 8883, pp. 189–196, 2014.

construct of triples. A query on a graph through SPARQL matches the pattern on the RDF graph and yields a set of data that matches the query. Assume a scenario where a set of SPARQL queries are to be operated upon an RDF. Though any permutation of a given set of SPARQL query is expected to generate identical final result, the time taken in processing individual query would vary depending on the sequence it follows within a permutation. This observation motivates us to explore the opportunity to reorder the SPARQL query sequences for speeding up the query processing.

Graphs, specifically RDFs have become critical for configuring and controlling the overwhelming volumes of information on the web. With the explosion of data into magnitude of Exascale and still growing, it is obvious that execution time of query on graphs will grow [5]. This issue certainly requires attention in terms of handling the query in a scalable form where large RDF graphs can be processed in reasonable time. The current state of art in SPARQL query optimization has few bodies of work for optimizing query executions [12], [8], [13]. These papers mainly focus either on speeding up the processing time by algorithmic approach on semantic optimization on queries or by taking advantage locality optimizations on split data across multiple computation nodes. In this paper, we focus on rather simpler method of reordering triple patterns for decreasing execution times of queries. Although there are several attempts to query reordering in SQL, these existing algorithms that are largely suitable and applied in the domain of SQL are unable to port in the similar form. Therefore, in our approach we design our algorithm exploiting SPARQL specific characteristics.

The rest of the paper is organized as follows: In section 2 we describe our approach to reorder the SPARQL query on RDF graph. In section 3 we describe the algorithm design and present our experimental result in section 4. Section 5 briefly describes the existing prior art and we conclude in section 6.

2   Approach

SPARQL is the standardized query language for RDF, the same way SQL is the standardized query language for relational databases [4]. RDF triples are comprised of a subject, predicate and object [3]. A SPARQL query consists of a set of triples or clauses where the subject, predicate and/or object can be variables or literals. Fig. 1 shows a sample query in SPARQL. The query has three clauses. The first clause has both subject and object as variables whereas the second and third clauses have a literal as their object. As we see here, predicate is a link between subject and object which describes the relationship between them. Our approach comes from the fact that the order in which query clauses are placed affects the response time of a query. We have analyzed the clauses on the basis of their structure and designed an algorithm to reorder them accordingly.

We observed that if clauses that generate less data are executed first, we get better response times. It is due to the fact that less data means less effort in joining data generated by two clauses. That means we can reduce cost of join operation by reordering queries such that clauses that generate less data are at the top. Now our aim is to find out the factors that affect the amount of data generated by query clauses. The amount of data generated depends on the number of exact matches between the query clauses and

```
SELECT ?student
  Where{
        ?student :takescourse ?course
        ?course :name  C1
        ?student :a ugstudent
  }
```

Fig. 1. A sample SPARQL query

graph data. If number of matches is less, less data will be generated. If we look very carefully, there are two factors which can affect the number of matches between queries and data.

1. Presence of literals in SPARQL query clauses
2. Frequency of predicate or links in the graph data.

2.1   Presence of Literals

```
?student   :takescourse   ?course
?student   :takescourse   C1
```

Fig. 2. A query to analyze

Consider the query clauses given in fig. 2. The first clause asks for all the matches where any student is taking any course. As both subject and object are variables, we only have to match the predicate in graph data. But in case of the second clause we have to match both predicate and object in graph data. The course has to be a specific course, i.e. C1. This reduces the amount of data generated. Although it depends on the kind of data that is required to be processed, it is possible that both of these clauses generate same output if we have only one course C1 in our graph data. But in general, we can say that presence of literals reduces the amount of data generated.

2.2   Frequency of Links

The number of times a link appears in graph data directly affects the number of exact matches between query clauses and data. It is self explanatory that if the number of triples in the graph data with a particular link is less, output data generated corresponding to that link will be less. Here also we can have cases where frequency of links does not affect data generated. There might be a case where there is no match for subject or predicate. In that case frequency is not significant. But for a general case, less frequency of link in the graph data means that less amount of data will be generated for that link.

3 Algorithm Design

The algorithm was designed keeping the presence of literals and the frequency of links in mind. Frequency of predicates(links) needs to be determined prior to execution of algorithm. It is stored in a file(freqfile).Queries are also stored in a file(queryfile).

A Java code was written to determine the frequency of links. The entire graph data was scanned and frequency of individual links was counted. Output was written to freqfile as mentioned above.

Algorithm 1 shows the important steps that follow. The inputs are queryfile and freqfile. Both files are read and query is ordered accordingly. Firstly, the clauses are arranged in increasing order of frequency and then clauses containing literals are brought at the to. As every variable in a query begins with the symbol '?', literals can be easily detected. Absence of the symbol indicates a literal.

Algorithm 1. Algorithm for Reordering Clauses in RDF

Require: $< key, val >$;
1: ReorderyQuery(freqfile, queryfile)
2: HashMap $< key, val >$
3: hm = new HashMap $< key, val >$();
4: read queryfile();
5: read freqfile();
6: **if** clause[i] contains link **then**
7:       hm.put(clause[i], freq[link]);
8:       sortbyvalue(hm);
9: **end if**
10: **for** $key \in$ hm **do**
11:       **if** key contains subject literal **then**
12:             val=0;
13:       **end if**
14:       **if** key contains object literal **then**
15:             val=1;
16:       **end if**
17:       **if** val=val+2 **then**
18:             sortbyvalue(hm);
19:       **end if**
20: **end for**
21: return hm;

More priority was given to literals. It means that a clause with two literals should be brought at top, followed by clauses with one literal and no literal. Thus, we get a reordered query with clauses arranged in increasing order of frequency. The clauses containing literals are at the top, but they are also ordered according to frequency.

We gave more priority to literals because clauses with literals are more likely to generate lesser amount of data. Of course we cannot be sure because it is possible that some link has so low frequency that it generates even lesser data than any clause containing literals. But as said, there are more chances that a clause containing literal will generate lesser data.

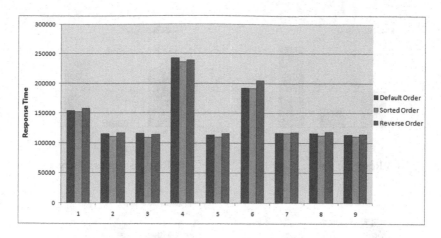

Fig. 3. Response times(in ms) for default, sorted and reverse order

4   Experimentation

4.1   SHARD Triple-Store

Scalable, High-Performance, Robust and Distributed (SHARD) [11] is a proof-of-concept high-performance, low-cost distributed computing technology to develop a highly scalable triple-store built on Hadoop and HDFS [1], [2]. SHARD triple-store persists data as standard RDF triples and runs queries over this data using the standard SPARQL query-language. It implements an approach called clause-iteration approach to run SPARQL queries over RDF data. Clause-Iteration algorithm utilizes MapReduce model for parallel execution of queries.// SHARD triple-store uses LUBM benchmark to generate test cases and evaluate the performance [7]. It gives us the response time for queries when run over generated graph data. We generated graph data that comprised of around three million nodes. We chose nine SPARQl queries from the test data of SHARD triple-store(or LUBM benchmark). Queries selected were simple with the number of clauses ranging from 2 to 5. We used SHARD triple-store for execution of queries and evaluation of our algorithm. Three machines were used. We first executed the queries without ordering them. After that, we reordered queries using our algorithm and then executed them again. We also reordered queries in an order that was completely reverse of the order given by our algorithm. We noted response time in each case.

The difference in response times for queries 1-9 can be seen in fig. 3. Results show that the order calculated by our algorithm reduced response times of the queries. Here default order refers to the original order of query clauses, sorted order is order given by the algorithm and reverse order is the reverse of sorted order. We do not see much difference here. This is because of the fact that there is a high difference between the response time of queries. So this graph does not have a suitable scale to show us the difference. If we analyse the results of each query separately we would be able to see the difference.

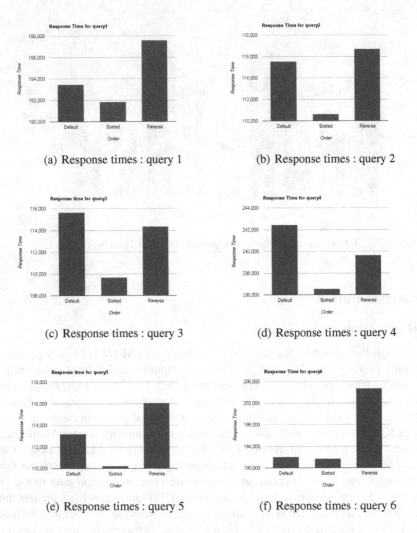

(a) Response times : query 1

(b) Response times : query 2

(c) Response times : query 3

(d) Response times : query 4

(e) Response times : query 5

(f) Response times : query 6

Fig. 4. Response times for queries:1-6

Form fig. 4 and 5, we can easily see the change in response times. The response time for all the queries is reduced. Also, note the difference between the response times of sorted order and reverse order. These results suggest that we can speed up querying over RDF data if we reorder them in the optimum order given by the algorithm.

5 Related Work

Query reordering is quite commonly done in SQL while querying relational databases. There are several attempts to query reordering in SQL [6]. However, in case of large RDF graph the focus is mostly in optimizing the SPARQL specific algorithmic changes

(a) Response times : query 7

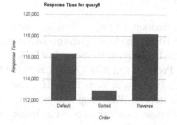
(b) Response times : query 8

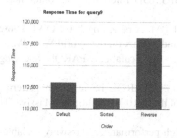
(c) Response times : query 9

Fig. 5. Response times for queries: 7,8,9

and partitioning across multiple nodes [12], [8]. There are few attempts to utilize MapReduce framework to improve the performance by exploiting parallelism in query processing. In [9] the authors exploit similarity found in a graph pattern and then iteratively process across multiple nodes. A work in similar line utilizing the large scale MapReduce framework is proposed in [10]. All these work while attempt to exploit the parallelism in the RDF data have very limited per node efficiency due to the fact that the storage layers (memory subsystem) is not optimize for the RDF data. In our approach we try to improve the performance by ordering the clauses that produces less data in the initial instances. This improves per node computation efficiency.

6   Conclusion and Future Work

In this paper our attempt is to show that query reordering has significant effect on response times. Our fairly simple approach of reordering clauses on RDF graph has resulted in 12% improvement on performance. It also helps in reducing the cost of join operation. Reordering clauses augmented with other performance enhancing methodologies, e.g., algorithmic and computational enhances would lead to better handling of large RDF graphs. In our future work we would explore to improve per node computation cost and overall joining operation to achieve better result.

References

1. Apache Hadoop, `http://hadoop.apache.org`
2. Apache HBase, `http://hbase.apache.org`
3. Resource Description Framework(RDF), `http://www.w3.org/RDF/`
4. Simple Protocol and RDF Query Language (SPARQL), `http://www.w3.org/TR/rdf-sparql-query`
5. Meeting report: Workshop on big data and extreme-scale computing, BDEC (2013), `http://www.exascale.org/bdec/sites/www.exascale.org.bdec/files/BDEC_Charleston_Workshop_Report_Final.pdf`
6. Goel, P., Iyer, B.: Sql query optimization: Reordering for a general class of queries. In: Proceedings of the 1996 ACM SIGMOD International Conference on Management of Data, SIGMOD 1996 (1996)
7. Guo, Y., Pan, Z., Heflin, J.: LUBM: A benchmark for OWL knowledge base systems. Web Semantics: Science, Services and Agents on the World Wide Web 3(2-3) (2005)
8. Huang, J., Abadi, D.J., Ren, K.: Scalable SPARQL querying of large RDF graphs. PVLDB 4(11) (2011)
9. Myung, J., Yeon, J., Lee, S.G.: SPARQL basic graph pattern processing with iterative MapReduce. In: Proceedings of the 2010 Workshop on Massive Data Analytics on the Cloud (2010)
10. Rohloff, K., Schantz, R.E.: High-performance, massively scalable distributed systems using the MapReduce Software Framework: The SHARD triple-store. In: Programming Support Innovations for Emerging Distributed Applications (2010)
11. Rohloff, K., Schantz, R.E.: Clause-iteration with mapreduce to scalably query datagraphs in the SHARD graph-store. In: Proceedings of the Fourth International Workshop on Data-intensive Distributed Computing, New York, NY, USA (2011)
12. Schmidt, M., Meier, M., Lausen, G.: Foundations of SPARQL query optimization. In: Proceedings of the 13th International Conference on Database Theory, New York, NY, USA (2010)
13. Zou, L., Chen, L., Özsu, M.T.: Distance-join: Pattern match query in a large graph database. Proc. VLDB Endow (2009)

Author Index